AF714739

INTRODUCTION TO ECOLOGY: INTERCONNECTIONS IN NATURE

INTRODUCTION TO ECOLOGY: INTERCONNECTIONS IN NATURE

Prof. Dr. Vinay Kumar Singh

Delve
PUBLISHING
www.delvepublishing.com

Introduction to Ecology: Interconnections in Nature

Prof. Dr. Vinay Kumar Singh

Delve Publishing
224 Shoreacres Road
Burlington, ON L7L 2H2
Canada
www.delvepublishing.com
Email: orders@arclereducation.com

© 2024 Delve Publishing

ISBN: 978-1-77469-950-8 (Hardcover)

This book contains information obtained from highly regarded resources. Reprinted material sources are indicated and copyright remains with the original owners. Copyright for images and other graphics remains with the original owners as indicated. A Wide variety of references are listed. Reasonable efforts have been made to publish reliable data. Authors or Editors or Publishers are not responsible for the accuracy of the information in the published chapters or consequences of their use. The publisher assumes no responsibility for any damage or grievance to the persons or property arising out of the use of any materials, instructions, methods or thoughts in the book. The authors or editors and the publisher have attempted to trace the copyright holders of all material reproduced in this publication and apologize to copyright holders if permission has not been obtained. If any copyright holder has not been acknowledged, please write to us so we may rectify.

Notice: Registered trademark of products or corporate names are used only for explanation and identification without intent of infringement.

Delve Publishing publishes wide variety of books and eBooks. For more information about Delve Publishing and its products, visit our website at www.delvepublishing.com.

ABOUT THE AUTHOR

Professor Vinay Kumar Singh, Ph. D., Department of Zoology, DDU Gorakhpur University, born on 15th July, 1969, has a brilliant academic carrier. Prof Singh has involved himself in research on Fasciolosis and its snail vectors control since 1993. Current research interests are Isolation, characterization and biological evaluation of Natural Plant Products and pest control, Vector-Borne Diseases, Toxicology & Pharmacology. Prof Singh has University teaching and research experiences of more than 30 years. Prof Singh has successfully isolated and identified pure molluscicidal compounds from more than a dozen medicinal plants and observed their mode of action on the snails at cellular level. His work is recognized internationally in the field of formulation and use of synthetic and bio-pesticides for integrated pest/ vector-diseases management. Prof Singh has devised a bioassay method for pesticides using isolated organs of snails. Currently, he is working on the pesticidal property of the different formulations of cow urine of Indian cow breeds singly or synergistically with bioactive product of plant molluscicides. The research publication on fasciolosis control by means of characterized plant derived pesticides, synergistic combinations, bait formulations or by reducing the reproductive capacity of vector/host to limit the population below threshold level which are widely quoted and acclaimed world over. Prof Singh has published 154 research papers in 94 leading International Journals of repute.

TABLE OF CONTENTS

LIST OF FIGURES

LIST OF ABBREVIATIONS

AaaS	Analytics as a Service
ABNJ	Areas Beyond National Jurisdiction
ADB	Asian Development Bank
AEM	Agri-Environment Measures
AIDS	Acquired Immune Deficiency Syndrome
AGGI	Annual Greenhouse Gas Index
ALRTI	Acute Lower Respiratory Infections
AMAP	Arctic Monitoring and Assessment Programme
AMCEN	African Ministerial Conference on the Environment
AMD	Acid Mine Drainage
AOC	Areas of Concern
ASEAN	Association of Southeast Asian Nations
ASGM	Artisanal and Small Scale Gold Mining
BACT	Best Available Control Technology
BaP	Benzo[a]pyrene
BAT	Best Available Techniques
BAU	Business-as-Usual
BC	Black Carbon
BECCS	Bioenergy with Carbon Capture and Storage
BEV	Battery Electric Vehicles
CA	Conservation Agriculture
CaCO3	Calcium Carbonate
CAP	Common Agricultural Policy (EU)
CAS	Chemical Abstract Service
CBMIS	Community-Based Monitoring and Information Systems
CBD	Convention on Biological Diversity

CBMIS	Community-Based Monitoring and Information Systems
CBO	Congressional Budget Office
CCAC	Climate and Clean Air Coalition for the Reduction of Short-Lived Climate Pollutants
CCAK	Clean Cookstoves Association of Kenya
CCAMLR	Conservation for the Antarctic Marine Living Resources
CCB	China Construction Bank Corporation
CCE	Climate Change Education
CCFSC	Central Committee for Flood and Storm Control
CCP	Command and Control Policies
CFC	Chlorofluorocarbon
CCS	Carbon Capture and Storage
CD	Compact Disk
CDC	Centre for Disease Control and Prevention (United States)
CEC	(1) Contaminants of Emerging Concern, or (2) Commission for Environmental Cooperation (under NAFTA)
CEDS	Community Emissions Data System
CENESTA	Centre for Sustainable Development and Environment
CFC-11	Trichlorofluoromethane
CH4	Methane
CITES	Convention on International Trade in Endangered Species of Wild Fauna and Flora
CLRTAP	Convention on Long-range Transboundary Air Pollution
CLTS	Community-Led Total Sanitation
CMM	Cutaneous Malignant Melanoma
CMS	Convention on the Conservation of Migratory Species of Wild Animals
CNS	Central Nervous System
CO	Carbon Monoxide
CO2	Carbon Dioxide
COMEAP	Committee on the Medical Effects of Air Pollutants
CONAFOR	National Forestry Commission of Mexico
COP	Conference of the Parties
COPD	Chronic Obstructive Pulmonary Disease

CORSIA	Carbon Offsetting and Reduction Scheme for International Aviation
CSDH	Commission on Social Detriments of Health
CSIRO	Commonwealth Scientific and Industrial Research Organisation (Australia)
CSML	Climate Science Model Language
CSO	Civil Society Organization
DaaS	Data as a Service
DALY	Disability Adjusted Life Year
DDT	Dichlorodiphenyltrichloro-Ethane
DEFRA	Department for Environment, Food and Rural Affairs (United Kingdom of Great Britain and Northern Ireland)
DESD	UN Decade of Education for Sustainable Development (UNCCD)
DFO	Department of Fisheries and Oceans (Canada)
DIF	Directory Interchanged Format
DLDD	Desertification, Land Degradation and Drought (UNCCD)
DPSIR	Drivers, Pressures, State, Impacts, Responses
DRR	Disaster Risk Reduction
DSI	Dust Storm Intensity
DSF	Desert Storm Frequency
DWAF	Department of Water Affairs and Forestry (South Africa)
EAP	Environmental Action Programme (EU)
EBA	Ecosystem-Based Adaptation
EBAFOSA	Ecosystem-Based Adaptation for Food Security Assembly
EC	European Commission
ECLAC	United Nations Economic Commission for Latin America and the Caribbean
EDC	Endocrine-Disrupting Chemicals
EEA	European Environment Agency
EGA	Environmental Goods Agreement
EIA	(1) Energy Information Administration (United States of America), or (2) Environmental Impact Assessment
EID	Emerging Infectious Diseases
EIP	Economic Incentive Policies

ELD	Economics of Land Degradation Initiative
EML	Ecological Metadata Language
ENGO	Environmental Non-Governmental Organizations
EPA	(1) Environmental Performance Assessment, or (2) Environmental Protection Agency (United States)
EPI	Environmental Policy Integration
ESA	(1) Environmentally Sensitive Area, or (2) European Space Agency
ESD	Education for Sustainable Development
ESDIS	Earth Science Data and Information System
EU	European Union
GLOBE	Global Learning and Observation to Benefit the Environment
GLRI	Great Lakes Restoration Initiative
GM	Genetically Modified
GMACC	Global Military Advisory Council on Climate Change
GMO	Genetically Modified Organism
GMSL	Global Mean Sea Level
GNH	Gross National Happiness
GPA	Global Program of Action for the Protection of the Marine Environment from Land-Based Activities
GRI	Global Reporting Initiative
GSP	Generalized Scheme of Preferences
$GtCO_2$	Gigatonne of Carbon Dioxide
FAO	Food and Agriculture Organization of the United Nations
FBS	Food Balance Sheet (FAO)
FBSP	Free Basic Sanitation Policy
FBWP	Free Basic Water Policy
FIT	Feed-in Tariff
FSC	Forest Stewardship Council
G7 United	Group of Seven (Canada, France, Germany, Italy, Japan, Kingdom, United States)
GACC	Global Alliance for Clean Cookstoves
GAEC	Good Agricultural and Environmental Conditions (EU)
GAP	Global Action Programme on Education for Sustainable Development
GAWSiS	Global Atmospheric Watch Station Information System

GBD	Global Burden of Disease
GBR	Great Barrier Reef
GBRMPA	Great Barrier Reef Marine Park Authority
GCM	(1) General Circulation model, or (2) Global Climate Model GDI Gasoline Direct Injection
GDP	Gross Domestic Product
GEA	Global Environmental Assessment
GEF	Global Environment Facility
GEMI	UN-Water's Integrated Monitoring Initiative
GEO	Global Environment Outlook
GEOSS	Global Earth Observation System of Systems
GESAMP	Joint Group of Experts on the Scientific Aspects of Marine Environmental Protection
GGEO	Global Gender and Environment Outlook
GGIM	Global Geospatial Information Management
GGW	Great Green Wall (China)
GHG	Greenhouse Gas
GIS	Geographical Information Systems
GLADIS	Global Land Degradation Information System
GLASOD	Global Assessment of Human-Induced Soil Degradation
GTP	Global Temperature Potential
GW	Gigawatt
GWG	Global Working Group
GWP	(1) Global Water Partnership, or (2) Global Warming Potential
ha	Hectares
HALE	Health Adjusted Life Expectancy
HCFC	Hydrochlorofluorocarbon
HEI	Health Effects Institute
HFC	Hydrofluorocarbon
Hg	Mercury
HIV	Human Immunodeficiency Virus
HLPF	High-level Political Forum on Sustainable Development
HS	Harmonized System
IAEG	(1) Inter-agency and Expert Group, or (2) Internationally Agreed Environmental Goals

IAM	Integrated Assessment Model
IAS	Invasive Alien Species
ICAO	International Civil Aviation Organization
ICCA	Indigenous and Community-Conserved Areas
ICCT	International Council on Clean Transportation
ICMM	International Council of Mining and Metals
ICP	International Cooperative Program
ICPDR	International Commission for the Protection of the Danube River
ICS	Improved Cookstove
ICT	Information and Communication Technology
IDMC	Internal Displacement Monitoring Centre
IDP's	Internally Displaced People
IEA	(1) International Energy Agency, or (2) Integrated Environmental Assessment
IFAW	International Fund for Animal Welfare
IFPRI	International Food Policy Research Institute
IGRAC	International Groundwater Resource Assessment Centre
IIED	International Institute for Environment and Development
IJC	International Joint Commission
IK	Indigenous Knowledge
ILBM	Integrated Lake Basin Management
ILO	International Labor Organization
IMO	International Maritime Organization
INBO	International Network of Basin Organizations
INDC	Intended Nationally Determined Contribution
INTERPOL	International Criminal Police Organization
IPLC	Indigenous Peoples and Local Communities
IPAT	Impact = Population x Affluence x Technology
IPBES	Intergovernmental Science-Policy Platformon Biodiversity and Ecosystem Services
IPCC	Intergovernmental Panel on Climate Change
IRDA	Iksandar Regional Development Authority
ISA	International Seabed Authority
ISO	International Organization for Standardization

ITF	International Transport Forum
ITPS	Intergovernmental Technical Panel on Soils
ITQ	Individual Transferable Quota
IUCN	International Union for the Conservation of Nature and Natural Resources
IUU	Illegal, Unreported and Unregulated Fishing
IWRM	Integrated Water Resources Management
JMP	Joint Monitoring Programme for Water Supply and Sanitation of WHO/UNICEF
JRC	Joint Research Centre (European Commission)
LA	Local Authorities
LAC	Latin America and the Caribbean
LAER	Lowest Achievable Emission Rate
LANCE	Land, Atmosphere Near Real-Time Capability for Earth Observing System
LCT	Low Carbon Technology
LDN	Land Degradation Neutrality (UNCCD)
LMO	Living Modified Organism
LMMA	Locally Managed Marine Areas
LPG	Liquefied Petroleum Gas
LPI	Living Planet Index
LSF	Large-Scale Fisheries
LTEMP	Glen Canyon Dam Long-Term Experimental and Management Plan
MAP	Mediterranean Action Plan for the Barcelona Convention
MARPOL	International Convention for the Prevention of Pollution from Ships
MCA	Minerals Council of Australia
MDG	Millennium Development Goal
MEA	(1) Multilateral Environmental Agreement, or (2) Millennium Ecosystem Assessment
MEB	Multiple Evidence Base
MPA	Marine Protected Area
MRV	Measurement, Reporting, and Verification
MSY	Maximizing Sustainable Yield

MTFR	Maximum Technologically Feasible Reduction
MUDP	Environmental Technology Development and Demonstration Program (Denmark)
MUFPP	Milan Urban Food Policy Pact
NAZCA	Non-State Actor Zone for Climate Action
NBS	Nature-Based Solutions
NBSAP	National Biodiversity Strategies and Action Plans
NcML	netCDF Markup Language
NCP	Nature's Contribution to People
NCSDs	National Councils for Sustainable Development
NDC	Nationally Determined Contribution
NDVI	Normalized Difference Vegetation Index
NEPA	National Environment Policy Act (United States)
NFCP	Natural Forest Conservation Program(China)
NGO	Non-Governmental Organization
NH_3	Ammonia
NIP	National Implementation Plan
NMVOC	non-methane volatile organic compounds
NO_2	Nitrogen Dioxide
NOAA	National Oceanic and Atmospheric Administration (United States)
NOWPAP	Action Plan for the Protection, Managementand Development of the Marine and Coastal Development of the Northwest Pacific Region
NOx	Nitrogen Oxides
NPP	Net Primary Productivity
NSPS	New Source Performance Standards
NT	No-Tillage
O3	Ozone
OC	Organic Carbon
OCP	Organochlorine Pesticides
ODA	Official Development Assistance
ODS	Ozone-Depleting Substance
OECD	Organisation for Economic Co-operatio nand Development
OSCAR	Observing Systems Capability Analysis and Review

OSPAR	Convention for the Protection of the Marine Environment of the North-East Atlantic
PA	Protected Area
PACC	Pacific Adaptation to Climate Change
PAME	Protected Area Management Effectiveness
PAH	Polycyclic Aromatic Hydrocarbons
PAR	Protect, Accommodate and Retreat
PAWS	Protection of Asian Wildlife Species
Pb	Lead
PBDE	Polybrominated Diphenyl Ethers
PBT	Persistent, Bioaccumulative Toxic Chemicals
PCB	Polychlorinated Biphenyls
PCFV	Partnership for Clean Fuel and Vehicles
PCN	Polychlorinated Naphthalene
PEER	Publishing and the Ecology of European Research
PES	Payment for Ecosystem Services
PFAS	Per- and Polyfluoroakyl Substances
PFC	Prefluorinated Chemicals
PFI	Port Fuel Injection
PHE	Phenanthrene
PM	Particulate Matter
PM10	Particulate Matter with a Diameter of 10
micrometres	0.01 Millimetre or Less
PM2.5	Particulate Matter with a Diameter of 2.5
micrometres	0.0025 Millimetre or Less
POPs	Persistent Organic Pollutants
PoWPA	Programme of Work on Protected Areas
PPP	(1) Purchasing Power Parity or, (2) Public Private Partnership
PPSR	Public Participation in Scientific Research
PRTR	Pollutant Release and Transfer Registers
R&D	Research and Development
RACT	Reasonably Available Control Technology
RBFM	Rights-Based Fisheries Management
RBG	Royal Botanic Garden, Kew

RBM	Resilience-Based Management
RBO	River Basin Organizations
RCP	Representative Concentration Pathways
RD&D	Research, Development and Demonstration
REACH	Registration, Evaluation, Authorisation and Restriction of Chemical (EU)
REC	Renewable Energy Credits
REDD	Reducing Emissions from Deforestation and Forest Degradation
RES	Renewable Energy Systems
RFMO	Regional Fisheries Management Organizations
RLI	Red List Index
ROD	Record of Decision
RPS	Renewable Portfolio Standard
SaaS	Software as a Service
SADC	Southern African Development Community
SAICM	Strategic Approach to International Chemicals Management
SBI	Subsidiary Body for Implementation
SBSTTA	Subsidiary Body on Scientific, Technical, and Technological Advice
SCBD	Secretariat of the Convention on Biological Diversity
SDG	Sustainable Development Goal
SDGIO	Sustainable Development Goals Interface Ontology
SDMX	Statistical Data and Metadata Exchange
SDS	Sand and Dust Storms
SDS-WAS	Sand and Dust Storm Warning Advisory and Assessment System
SEA	Strategic Environmental Assessment
SEEA	System of Environmental-Economic Accounting
SensorML	Sensor Model Language
SFA	State Forestry Administration (China)
UNEP-UNECE	United Nations Environment Programme – Economic Commission for Europe
UNEP-WCMC	United Nations Environment Programme – World Conservation Monitoring Centre

UNESCAP	United Nations Economic and Social Commission for Asia and the Pacific
UNESCO	United Nations Educational, Scientific and Cultural Organization
UNFCCC	United Nations Framework Convention on Climate Change
UNFPA	United Nations Population Fund
UNGA	United Nations General Assembly
UNHCR	United Nations High Commissioner for Refugees
UNICEF	United Nations International Children's Emergency Fund
UNIDO	United Nations Industrial Development Organization
UNISDR	United Nations Office for Disaster Risk Reduction
UNODC	UN Office on Drugs and Crime
UNSCN	United Nations System Standing Committee on Nutrition
UNSDSN	United Nations Sustainable Development Solutions Network
US EPA	United States Environmental Protection Agency
USA	United States of America
USAID	United States Agency for International Development
UV	Ultraviolet
VGGT	FAO Voluntary Guidelines on the Responsible Governance of Tenure
VME	Vulnerable Marine Ecosystems
VOC	Volatile Organic Compounds
VWE	Vulnerable Marine Ecosystems
WAD	World Atlas of Desertification
WASH	Water, Sanitation and Hygiene
WAVES	Wealth Accounting and the Valuation of Ecosystem Services
WBCSD	World Business Council for Sustainable Development
WDPA	World Database on Protected Areas
WEF	World Economic Forum
WFD	Waste Framework Directive of the EU
WfW	Working for Water programme
WHC	World Heritage Convention
WHO	World Health Organization
WIPO	World Intellectual Property Organization

WMO	World Meteorological Organization
WOCAT	World Overview of Conservation Approaches and Technologies
WRI	World Resources Institute
WTO	World Trade Organization
WWAP	World Water Assessment Programme
WWF	World Wide Fund for Nature
WWF-SA	World Wide Fund for Nature, South Africa
ZSL	Zoological Society of London

PREFACE

Ecology is the scientific study of the relationships between organisms and their environment. It is a field that seeks to unravel the intricate web of interactions that define life on Earth. At its core, ecology delves into the interconnections in nature, exploring how different species, populations, and ecosystems interact and influence one another. In this essay, we will embark on a comprehensive journey into the world of ecology, examining its fundamental principles, the importance of understanding interconnections in nature, and the applications of ecological knowledge in solving real-world problems.

Ecology is built upon a set of fundamental principles that serve as the foundation for understanding the complex interactions that occur in natural systems. These principles are crucial for unraveling the mysteries of life on Earth:

- *Interdependence:* The cornerstone of ecology is the recognition that all living organisms are interconnected. Every species, from the tiniest microbe to the largest whale, depends on other organisms and their environment for survival. This interdependence forms the basis of ecological relationships.
- *Levels of Organization:* Ecology studies life at various levels of organization, ranging from individuals to populations, communities, ecosystems, and the biosphere. Understanding these levels helps ecologists analyze how different factors influence life on Earth.
- *Energy Flow:* All living organisms require energy to carry out their life processes. Energy flows through ecosystems in a one-way path, starting with the sun, which provides energy for photosynthesis, and eventually dissipates as heat. This energy flow underpins the functioning of ecosystems.
- *Nutrient Cycling:* Nutrients, such as carbon, nitrogen, and phosphorus, are essential for life. Ecology explores how these nutrients cycle through ecosystems, from the soil to plants, animals, and decomposers, ensuring a continuous supply for living organisms.
- *Adaptation and Evolution:* Organisms evolve over time through the process of natural selection. Ecology examines how adaptations to specific environments enable species to thrive and how they may shape ecosystems.
- *Succession:* Ecological communities change over time through a process known as ecological succession. This principle explores how ecosystems evolve and adapt after disturbances, such as fires or deforestation.

The study of interconnections in nature is paramount for several reasons:

- *Biodiversity Conservation:* Understanding the complex web of interactions among species and their environment is essential for conserving biodiversity. By recognizing the role of each species in an ecosystem, we can develop strategies to protect endangered species and ecosystems.
- *Ecosystem Services:* Ecosystems provide a multitude of services, including clean air and water, pollination of crops, and climate regulation. Recognizing the interdependencies within ecosystems helps us appreciate the value of these services and the consequences of disrupting them.
- *Human Well-Being:* Human societies are intimately linked to the natural world. Knowledge of ecological interconnections aids in sustainable resource management, reducing the negative impacts of human activities on the environment, and ensuring the well-being of future generations.
- *Climate Change Mitigation:* Ecological research informs our understanding of climate change and helps develop strategies for mitigating its effects. For instance, forests act as carbon sinks, absorbing carbon dioxide from the atmosphere, which is crucial for climate stabilization.

Ecological knowledge has practical applications in addressing real-world challenges:

- *Conservation Biology:* Ecologists work to preserve endangered species and protect biodiversity. Their research informs conservation strategies, such as habitat restoration and captive breeding programs.
- *Agriculture and Pest Control:* Understanding ecological interactions in agricultural systems is essential for sustainable farming practices and effective pest management, reducing the need for harmful pesticides.
- *Urban Planning:* Ecological principles guide urban planners in designing cities that are more sustainable and resilient to environmental changes. Concepts like green infrastructure and urban biodiversity enhance the quality of life for city dwellers.
- *Public Health:* Ecology plays a role in tracking disease vectors, understanding the spread of diseases, and mitigating public health risks. For example, understanding the ecology of mosquitoes helps in the control of vector-borne diseases like malaria and Zika virus.
- *Ecosystem Restoration:* Ecological principles are applied to restore damaged ecosystems, such as wetlands, forests, and coral reefs. Restoration efforts help reverse habitat degradation and enhance ecological resilience.

Ecology is a multidisciplinary science that unravels the intricate tapestry of life on Earth. By studying the interconnections in nature, ecologists provide invaluable insights into the functioning of ecosystems, the conservation of biodiversity, and the solutions to pressing environmental challenges. As we face a future marked by climate change and environmental degradation, the knowledge gained from ecology is more critical than ever. By appreciating the profound interdependence of all life forms, we can work towards a harmonious coexistence with nature, ensuring the health and sustainability of our planet for generations to come.

CHAPTER 1

Meaning and Definition of Ecology

CONTENTS

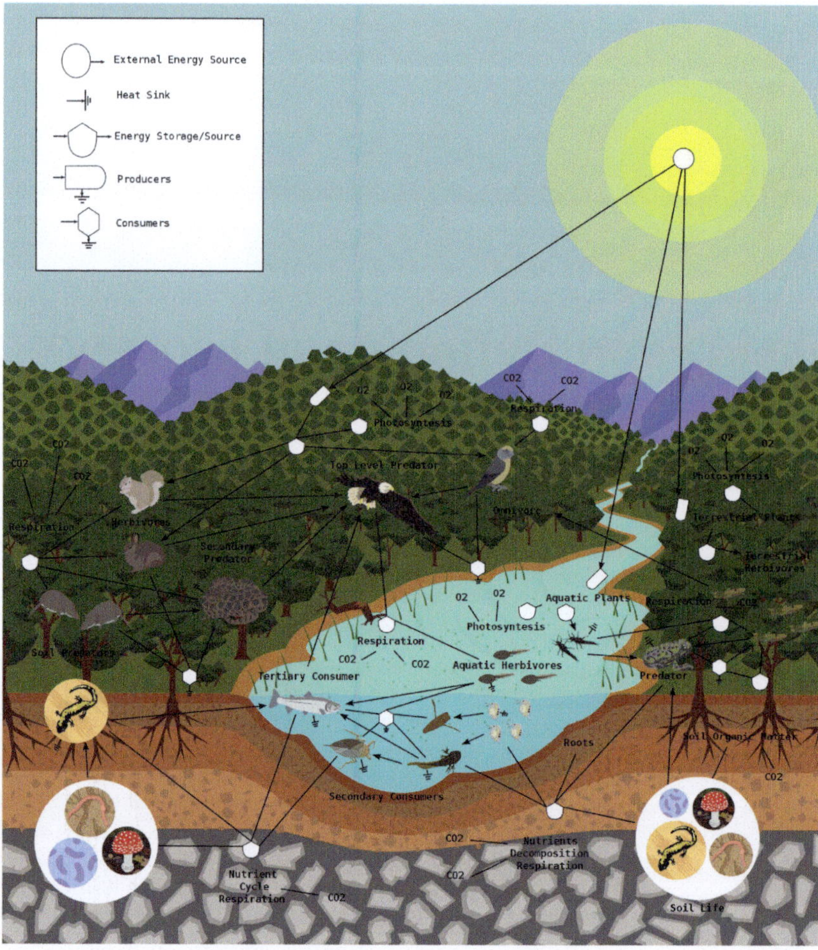

Figure 1. Ecology: Interconnections in Nature and Society

(Source: By Original: Thompsma Vector: Pixelsquid – This file was derived from: FoodWeb.jpg: by Thompsma, CC0, https://commons.wikimedia.org/w/index.php?curid=102527971, Creative Commons License)

1.1. INTRODUCTION

Ecology, a branch of biology, is the scientific study of the interactions between organisms and their environment. It seeks to understand the intricate web of relationships that shape the distribution and abundance of life on Earth. At its core, ecology is a holistic science, focusing not only on individual organisms but also on their interactions, populations, communities, ecosystems, and the biosphere.

1.1.1. Levels of Ecological Organization

Ecology investigates life across various levels of organization:

- Individuals: At the smallest scale, ecologists study how individual organisms adapt to their surroundings and interact with other individuals of the same or different species.
- Populations: Populations consist of individuals of the same species living in the same area. Ecologists explore factors affecting population size, growth, and dynamics.
- Communities: A community comprises multiple populations of different species living and interacting within the same area. Community ecology investigates species interactions like predation, competition, and mutualism.
- Ecosystems: Ecosystems encompass both living organisms and their physical environment. They include elements such as soil, water, nutrients, and climate. The flow of energy and cycling of matter within ecosystems are central ecological concepts.
- Biosphere: The biosphere represents the sum of all Earth's ecosystems. It involves studying how life, from the deepest ocean trench to the highest mountaintop, is interconnected and influences global processes.

1.1.2. Key Ecological Concepts

- Energy Flow: Energy from the sun fuels life on Earth. Through photosynthesis, plants and other autotrophs convert solar energy into chemical energy, which is then passed through food webs as heterotrophs (consumers) eat plants or other animals. Energy flow is unidirectional, eventually dissipating as heat.
- Nutrient Cycling: Essential nutrients like carbon, nitrogen, and phosphorus are recycled through ecosystems. Decomposers break down dead organic matter, returning nutrients to the soil or water, where they can be taken up again by plants.
- Adaptation and Evolution: Organisms evolve over time through natural selection, adapting to their environments. Ecologists study how these adaptations affect survival, reproduction, and species diversity.
- Succession: Ecological succession examines how ecosystems

change over time in response to disturbances, such as fires or volcanic eruptions. Primary succession occurs on bare substrates, while secondary succession follows disturbances that leave some remnants of the previous community.

1.1.3. Applied Aspects of Ecology

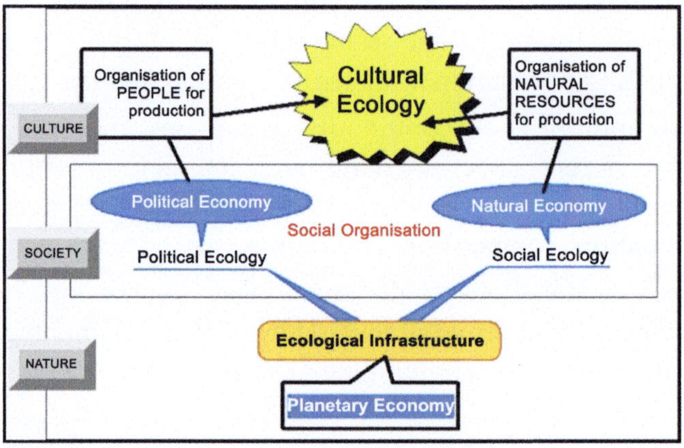

Figure 2. Applied Ecology

(Source: Wikibooks, Creative Commons License)

Ecology has profound real-world applications:

- Conservation Biology: Ecologists play a crucial role in conserving biodiversity. They assess the status of endangered species and design conservation plans, often involving habitat restoration and captive breeding programs.
- Agriculture and Pest Management: Understanding ecological interactions helps develop sustainable farming practices and integrated pest management strategies, reducing the need for chemical pesticides.
- Urban Ecology: Urban planners use ecological principles to design sustainable, green cities. Concepts like green roofs, urban forests, and sustainable transportation systems improve urban environments.
- Public Health: Ecology informs disease ecology and epidemiology. Ecologists study disease vectors, transmission dynamics, and the impacts of environmental changes on public health.
- Climate Change Mitigation: Ecological research contributes to

climate change mitigation by assessing carbon sequestration in forests, the impact of climate change on ecosystems, and strategies for adaptation and mitigation.

- Ecosystem Restoration: Ecologists lead efforts to restore damaged ecosystems, enhancing their resilience and functionality. This is vital for conserving natural resources and promoting ecosystem services.

Ecology is the science that uncovers the intricate connections between living organisms and their environment. It provides both a scientific understanding of the natural world and practical tools for addressing critical environmental challenges. The applications of ecological knowledge are essential for promoting sustainability, conserving biodiversity, and ensuring the well-being of both ecosystems and human societies in an increasingly interconnected and changing world.

1.2. MEANING OF ECOLOGY

Figure 3. Meaning of Ecology

(Source: Wikipedia.org, Creative Commons License)

Ecology is a multifaceted term with various meanings and interpretations, depending on the context in which it is used. Here, we explore the multiple meanings of ecology in great detail:

Scientific Discipline:

- Ecology as a Scientific Field: Ecology is primarily known as a scientific discipline that studies the interactions between living organisms and their environments. Ecologists investigate the relationships among species, the flow of energy and nutrients in ecosystems, and the impact of environmental factors on the distribution and abundance of organisms. This scientific approach to ecology seeks to understand the natural world's complexity and dynamics.

Environmental Concerns:

- Ecology as Environmental Concern: Ecology can also refer to the broader awareness and concern for the environment and the need to protect and preserve it. In this sense, it signifies a societal commitment to sustainable practices, conservation, and environmental stewardship.

Relationships and Interactions:

- Ecology as Relationships: Ecology, in a more general sense, refers to the relationships and interactions between organisms and their surroundings. It encompasses the interconnectedness of all living things and the ways in which they depend on each other and the environment.

Human Interactions with Nature:

- Human Ecology: Human ecology is a branch of ecology that focuses on the interactions between humans and their environment. It examines how human activities and behaviors impact ecosystems and how these interactions, in turn, affect human societies.

Design and Planning:

- Ecology in Design and Planning: In architecture and urban planning, ecology is often used to describe the incorporation of natural elements and sustainable principles into the design of buildings, landscapes, and cities. Ecological design aims to create environments that harmonize with nature and minimize negative impacts on the ecosystem.

Philosophical and Metaphorical Usage:

- Ecology as a Metaphor: In a metaphorical or philosophical sense, ecology can refer to the interconnectedness and balance of various elements or systems. For example, the "ecology of the mind" might describe the interplay of thoughts and ideas within a person's consciousness, highlighting the need for harmony and equilibrium.

Economic and Business Contexts:

- Business Ecology: This term is used in business and economics to describe the study of the interactions and relationships within an industry, market, or economic system. It emphasizes the need for a sustainable, balanced approach to economic activities that considers environmental and social impacts.

Computer Science and Information Systems:

- Ecology in Computer Science: In computer science and information systems, the concept of ecology can be applied to describe the relationships and interactions within complex software systems, networks, or data ecosystems. It signifies the need for understanding and managing the dynamics of interconnected digital systems.

Political and Social Movements:

- Social and Political Ecology: These terms describe fields of study and social movements that examine the interconnections between politics, society, and the environment. They explore how political and economic systems impact natural resources, sustainability, and social justice.

Education and Pedagogy:

- Ecological Education: Ecological education emphasizes the importance of teaching and learning about ecology and environmental issues. It promotes an understanding of the natural world and encourages responsible environmental behavior.

Art and Culture:

- Ecological Art and Culture: In the realm of art and culture, ecological themes and motifs are often used to convey messages about environmental consciousness, sustainability, and the relationship between humanity and the natural world.

Ecology encompasses a wide range of meanings and applications, from its foundational role as a scientific discipline that studies the natural world's interconnections to its broader implications for environmental awareness, human interactions with nature, design and planning, and its metaphorical use in various contexts. Understanding these diverse meanings of ecology helps us appreciate its significance in addressing environmental challenges and promoting a harmonious relationship between humans and the planet.

1.3. DEFINITIONS OF ECOLOGY

Ecology is a multidisciplinary field that has evolved over time, and different scientists have contributed to its development by providing various definitions and perspectives. Here are detailed definitions of ecology by some prominent ecologists and scientists:

- Ernst Haeckel (1866): Ernst Haeckel, a German biologist, is often credited with coining the term "ecology." In his work, he defined ecology as the "science of relations between organisms and their environment."
- Eugene P. Odum (1963): Eugene P. Odum, an influential American ecologist, provided a comprehensive definition: "Ecology is the scientific study of the structure and function of nature, including organisms (individuals, populations, and communities), the abiotic environment (such as climate, soil, and water), and the interactions between them."
- G. Evelyn Hutchinson (1947): Hutchinson, an ecologist known for his work on ecological niches, defined ecology as "the branch of biology dealing with the relations and interactions between organisms and their environment, including other organisms."
- Charles J. Krebs (2009): Charles J. Krebs, in his book "Ecology: The Experimental Analysis of Distribution and Abundance," defines ecology as "the scientific natural history of the Earth."
- Howard T. Odum (1971): Howard T. Odum, a pioneer in ecosystem ecology, defined ecology as "the study of systems that include humans and nature as an integral part."
- Barry Commoner (1971): Barry Commoner, an ecologist and environmentalist, defined ecology as "the science of the total environment." He emphasized the interconnectedness of all aspects of the environment.
- E.O. Wilson (1984): Edward O. Wilson, a prominent biologist and proponent of biodiversity conservation, defined ecology as "the study of the relationships between organisms and all aspects of their environment."
- Robert E. Ricklefs (1979): Robert E. Ricklefs, in his book "Ecology," defines ecology as "the scientific study of the distribution and abundance of organisms and their interactions with their environment."

- Daniel Botkin (1995): Daniel Botkin, an ecologist and environmental scientist, defined ecology as "the study of how organisms interact with each other and their physical environment and how these interactions change over time."
- Ramón Margalef (1997): Ramón Margalef, a Spanish ecologist, defined ecology as "the study of ecosystems and their functioning, including the processes that regulate the composition, structure, and dynamics of these systems."

These definitions reflect the evolving nature of ecology as a scientific discipline that encompasses a wide range of topics, from individual organisms to entire ecosystems. Ecologists have recognized the intricate web of interactions between organisms and their environment and the importance of understanding these relationships for addressing environmental challenges and conservation efforts.

The most widely accepted definition of ecology today, which captures the essence of the field in great detail, can be expressed as follows:

"Ecology is the scientific study of the relationships and interactions between living organisms and their environment. It seeks to understand the distribution and abundance of organisms, the dynamics of ecosystems, and the processes that influence the structure and function of natural systems. Ecology encompasses multiple levels of organization, from individual organisms to populations, communities, ecosystems, and the biosphere. It examines the flow of energy, the cycling of nutrients, and the influence of biotic and abiotic factors on the diversity and adaptation of life forms. Ecology plays a crucial role in addressing environmental challenges, conserving biodiversity, and promoting sustainable practices to ensure the well-being of both ecosystems and human societies in an interconnected world."

This comprehensive definition highlights several key aspects of ecology:

- Scientific Study: Ecology is a rigorous and systematic scientific discipline that employs empirical observation, experimentation, and data analysis to investigate natural systems. It relies on the scientific method to formulate hypotheses, conduct research, and draw evidence-based conclusions.
- Relationships and Interactions: Ecology emphasizes the examination of interactions among living organisms and their surroundings. These interactions include competition, predation,

mutualism, and the influence of physical and chemical factors in the environment.

- Scale and Hierarchy: Ecologists study ecological phenomena at multiple levels of organization, from the smallest scale (individual organisms) to the largest scale (the biosphere). This hierarchical approach allows for a comprehensive understanding of ecological processes.
- Ecosystem Dynamics: Ecology explores the dynamics of ecosystems, including the cycling of energy and nutrients, succession, and response to disturbances. It considers both the biotic (living) and abiotic (non-living) components that shape ecosystems.
- Adaptation and Diversity: The field of ecology delves into how organisms adapt to their environments and the role of diversity in maintaining the resilience and functionality of ecosystems. It examines how natural selection drives evolutionary processes.
- Interconnectedness: Ecology recognizes that life on Earth is intricately interconnected. Changes in one part of an ecosystem can have ripple effects throughout the system and even influence global processes.
- Applied Aspects: Ecology has practical applications in addressing environmental challenges, conserving biodiversity, and promoting sustainability. It informs strategies for managing natural resources, mitigating climate change, and protecting ecosystems.
- Holistic Perspective: Ecology takes a holistic approach to understanding the natural world, emphasizing the need to consider all relevant factors and interactions when studying ecological systems.

In summary Mary, the most accepted definition of ecology today underscores its role as a scientific discipline that investigates the relationships between organisms and their environment across various levels of organization. It is a vital field of study for addressing pressing environmental issues and ensuring the coexistence of diverse life forms on Earth while maintaining the health and functionality of ecosystems.

1.4. REFERENCES

1. Adams, J. (1986). 'Marine life of the Isca Silurum Tunnels, Doolin Point, Co. Clare, S. W. Ireland,' Isca Caving Club Journal, 7, 13–14.
2. An Foras Forbartha (1972). 'An investigation into the effects of fertilizer application on limestone grassland in the Burren, Co. Clare,' Dublin
3. Anon. (1978). 'Corrigendum: The Status of Mercurialis perennis in the Burren,' The Irish Naturalists' Journal, 19(7), 256.
4. Bates, M. A. (1988). A Review of the management and its effects on the vegetation in the Burren national park, Co. Clare. Unpublished Report to the Office of Public Works. Farnborough College of Technology.
5. Bonham, F. R. H. & Fairley, J. S. (1984). 'Observations on a Herd of Feral Goats Capra (Domestic) in the Burren,' The Irish Naturalists' Journal, 21(5), 208–212.
6. British Ecological Society (1962). 'The Burren and its problems: Summary of a paper by Prof. D. A. Webb,' Journal of Ecology, 50, Oxford: Blackwell Scientific Publications, 846–848.
7. Bunce, C. (1997). 'Caddis fly larval tubes in Polltalloon,' Irish Speleology, 16, 24.
8. Byrne, R. A. (1981). Ecological comparisons of three water bodies in The Burren District of County Clare. Unpublished MSc thesis No. 436. Trinity College Dublin.
9. Byrne, R. A. & Reynolds, J. D. (1982). 'Preliminary notes on a doline in Lough Gealain, the Burren, County Clare,' The Irish Naturalists' Journal, 20(9), 375–377.
10. Cabot, D. (1999). 'The Burren and Turloughs,' in Cabot, D. Ireland: A Natural History, London: Harper Collins, 191–230.
11. Corry, T. H. (1880). 'Notes of a botanical ramble in the County of Clare, Ireland,' Proceedings of the Belfast Natural History and Philosophical Society, 167–207.
12. Curtis, T. G. F. (1981). 'A further station for Mercurialis perennis, L. in the Burren with comments on its status there,' The Irish Naturalists' Journal, 20(5), 184–185.
13. Curtis, T. G. F. & McGough, H. N. (1981). A survey of the wetlands of the Fergus catchment and adjoining areas. FWS internal report.

14. D'Arcy, G. & Hayward, J. (1992). The Natural History of the Burren. London: Immel Publishing.
15. D'Auria, L. (1993). The impact of tourists on the Burren vegetation on County Clare, Ireland. Unpublished BSc thesis. Dublin: University College Dublin.
16. David, R. (1960). 'The Burren Flora,' The Irish Naturalists' Journal, 13(7), 169.
17. de Válera, M. (1960). 'Interesting Seaweeds from the Shores of the Burren,' The Irish Naturalists' Journal, 13(7), 168.
18. De Valera, M. & Cooke, J. (1979). 'Seaweed in Burren Grykes,' The Irish Naturalists' Journal, 19(12), 436.
19. de Valéra, M., Pybus, C., Casley, B. & Webster, A. (1979). 'Littoral and Benthic Investigations on the West Coast of Ireland, X. Marine Algae of the Northern Shores of the Burren, Co. Clare,' Proceedings of the Royal Irish Academy. Section B: Biological, Geological, & Chemical Science, 79, 259–269.
20. Deenihan, A., Donlan, J., Breen, J. & Moles, R. (2009). 'Short Communication: Mid-Term Impacts of Excluding Large Grazing Animals on a Burren Grass/Scrubland Patch,' Biology and Environment: Proceedings of the Royal Irish Academy, 109B (2), 107–113.
21. Dickinson, C. H. & Thorp, T. K. (1968). 'Epiphytic Lichens on Corylus avellana in the Burren, County Clare,' Lichenologist, 4, Oxford: Blackwell Scientific Publications, 66–72.
22. Dickinson, C. H., Pearson, M. C. & Webb, D. A. (1964). 'Some Micro-Habitats of the Burren, Their Micro-Environments and Vegetation,' Proceedings of the Royal Irish Academy. Section B: Biological, Geological, & Chemical Science, 63, 291–302.
23. Doyle, G. J. (1986). 'Lychnis flos-cuculi, L., & Carex disticha Hudson in the Burren Hills,' The Irish Naturalists' Journal, 22(2), 74–77.
24. Doyle, G. J. (1993). 'Cuscuta epithymum (L.) L. (Convolvulaceae), Its Hosts and Associated Vegetation in a Limestone Pavement Habitat in the Burren Lowlands in County Clare (H9), Western Ireland,' Biology and Environment: Proceedings of the Royal Irish Academy, 93B (2), 61–67.
25. Drew, D. & Magee, E. (1994). 'Environmental Implications of Land Reclamation in the Burren, Co. Clare: a preliminary analysis,' Irish Geography, 27(2), 81–96.

26. Dunne, F. (1994). Sampling and establishment of permanent vegetation quadrats in the Burren National park. Unpublished Report to the Office of Public Works.
27. Fairley, J. S. (1967). 'The Fieldmouse Apodemus sylvaticus (L.) in the Burren,' The Irish Naturalists' Journal, 15(11), 330–331.
28. Fairley, J. S. & O'Gorman, F. (1974). 'Food of Pine Martens in the Burren,' The Irish Naturalists' Journal, 18(4), 125.
29. Fanning, M. (1995). The Ecology of Limestone heaths in the Burren, Co. Clare. Unpublished BSC thesis. Dublin: University College Dublin, Department of Botany.
30. Fanning, M. (1995). Vegetation studies in The Burren National Park. Unpublished Report.
31. Fanning, M. (1996). Description of the soil types and their related vegetation types in the Burren National Park. Unpublished Report.
32. Foot, F. J. (1863). 'Natural History Notes on the Mammalia of the West coast of Clare,' Proceedings of the Natural History Society of Dublin, 3, 104–106.
33. Foot, F. J. (1864). 'On the distribution of plants in the Burren, County of Clare,' Proceedings of the Royal Irish Academy, 24, 143–160.
34. Foot, F. J. (1871). 'On the Distribution of Plants in Burren, County of Clare,' The Transactions of the Royal Irish Academy, 24, 143–160.
35. Forbes, R. S. (1989). 'The Population Behaviour of Pyramidal Bugle Ajuga pyramidalis, L. in the Burren, Co Clare,' The Irish Naturalists' Journal, 23(2), 54–59.
36. Gallagher, R. N. & Fairley, J. S. (1979). 'A Population Study of Fieldmice Apodemus sylvaticus in the Burren,' Proceedings of the Royal Irish Academy. Section B: Biological, Geological, & Chemical Science, 79, 123–137.
37. Goodwillie, R. (1999). 'Campanula trachelium, L. in Clare (H9), New to the Burren,' The Irish Naturalists' Journal, 26(7/8), 286.
38. Gough, K. (1952). 'Orchis traunsteinerioides Pugsl. in Clare,' The Irish Naturalists' Journal, 10, 273.
39. Grierson, H. (1902). 'Some land and freshwater shells from Co. Clare,' Irish Naturalist, 11, 139–140.
40. Harding, T. (1968). 'A List of Woodlice (Crustacea; Isopoda: Oniscoidea) from the Burren, Co. Clare,' The Irish Naturalists' Journal, 16(1), 16–17.

41. Harrington, T. J. (1996). 'Observations on the Macrofungi of Dryas Communities in the Burren, Western Ireland,' The Irish Naturalists' Journal, 25(5), 169–174.
42. Harrington, T. J. (2003). 'Relationships between Macrofungi and Vegetation in the Burren,' Biology and Environment: Proceedings of the Royal Irish Academy, 103B (3): Understanding the Burren, 147–159.
43. Heslop-Harrison, J. (1949). 'Pinguicula grandiflora (Lam.) in north Clare,' Irish Naturalists' Journal, 9, 311.
44. Heslop-Harrison, J. (1961). 'A Note on Temperature and Vapour Pressure Deficit under Drought Conditions in Some Microhabitats of the Burren Limestone, Co. Clare,' Proceedings of the Royal Irish Academy. Section B: Biological, Geological, & Chemical Science, 61, 109–114.
45. Heslop-Harrison, J., Wilkins, D. A. & Greene, S. W. (1960). 'Arenaria norvegica (Gunn), a species new to the Irish Flora, in County Clare,' Irish Naturalists' Journal, 13, 267–268.
46. Ivimey-Cook, R. B. (1965). 'The vegetation of solution cups in the limestone of the Burren Co. Clare,' The Journal of Ecology, 53, Oxford: Blackwell Scientific Publications, 437–445.
47. Ivimey-Cook, R. B. & Proctor, M. C. F. (1966). 'Plant Communities of the Burren, Co. Clare,' Proceedings of The Royal Irish Academy, 64B, 211–301.
48. Jeffrey, D. W. (1995). 'The management of Burren Grasslands,' in Jeffrey, D. W., Jones, M. B. & McAdam, J. H. (eds.) Irish grasslands – their biology and management. Dublin: Royal Irish Academy, 267–75.
49. Jeffrey, D. W. (2003). 'Grasslands and Heath: a Review and hypothesis to Explain the Distribution of Burren Plant Communities,' Biology & Environment: Proceedings of the Royal Irish Academy, 103(3), 111–123.
50. Jelicic, Lj. & O'Connell, M. (1992). 'History of vegetation and land use from 3200 B. to the present in the north-west Burren, a karstic region of western Ireland,' Vegetation History and Archaeobotany, 1, 119–140.
51. Keane, S. (1990). Vegetation and land use studies in the Eastern area of the Burren, Co. Clare. Unpublished MSc Thesis. University College Galway, Department of Botany.

52. Keane, S. & Sheehy Skeffington, M. (1995). 'Vegetation in the East Burren Uplands in relation to land use and conservation,' in Jeffrey, D. W., Jones, M. B. & McAdam, J. H. (Eds.) Irish grasslands- their biology and management, Dublin: Royal Irish Academy, 267–75.

53. Kelly–Quinn, M., Bradley, C., Murray, D., Tierney, D., Ashe, P., Bracken, J. & McGarrigle, M. (2003). 'Physico–Chemical Characteristics and macroinvertebrate Communities of the Caher River,' Biology & Environment: Proceedings of the Royal Irish Academy,103 (3), 187–196.

54. Kelly, D. L. (1980). Mullaghmore Wood, Burren, Co. Clare. Unpublished Report.

55. Kelly, D. L. & Kirby, E. N. (1982). 'Irish Native Woodlands over Limestone,' in White, J. (ed.) Studies on Irish Vegetation, Dublin: Royal Dublin Society, 181–198.

56. Kimberley, S., & Waldren, S. (2012). 'Examinations of turlough soil property spatial variation in a conservation assessment context,' Biology & Environment: Proceedings of the Royal Irish Academy, 112(2), 193–205.

57. Kirby, E. N. (1980). 'New additions to the lichen flora of the Burren,' The Irish Naturalists' Journal, 20(1), 45.

58. Kirby, E. N. (1981). An Ecological and Phytosociological study of Corylus avellana, L. in the Burren. Unpublished PhD thesis No. 985. University College Galway.

59. Kirby, E. N. & MacGowran, B. A. (1979). A Vegetation survey of the proposed National Park area in The Burren, Co. Clare. Unpublished Report to the Office of Public Works. University College Galway, Dept. of Botany.

60. Lambe, E., Mitchell, M. E. & O'Connell, M. (1978). 'A reassessment of the status of Mercurialis perennis, L. in the Burren, Co. Clare,' The Irish Naturalists' Journal, 19(5), 157–158.

61. Lansbury, I. (1966). 'Notes on the Hemiptera, Coleoptera, Diptera and Other Invertebrates of the Burren, Co. Clare and Inishmore, Aran Islands,' Proceedings of the Royal Irish Academy. Section B: Biological, Geological, & Chemical Science, 64, 89–115.

62. Levinge, H. C. (1892). 'Neotinea intacta in County Clare,' Journal of Botany, British and Foreign, 30, 194–195.

63. Liston, M. & Harrington, T. (2012). 'Mycorrhizal Associations

of Arctostaphylos Uva-Ursi and Helianthemum Oelandicum spp Piloselloides in the Burren Dryas Heaths,' Biology & Environment: Proceedings of the Royal Irish Academy, 112(3), 311–313.

64. Lynn, D. E. & Waldren, S. (2003). 'The use of Ranunculus repens as an Indicator Species for Assessing the Extent of flooding in Turlough Basins,' Biology & Environment: Proceedings of the Royal Irish Academy, 103(3), 161–168.
65. Malloch, A. J. C. (1976). 'An Annotated Bibliography of the Burren,' Journal of Ecology, 64(3), 1093–1105.
66. McCarthy, M. (1983). 'The composition of some calcicolous lichen communities in the Burren, Western Ireland,' Lichenologist 15(3), London: Academic Press, 231–248.
67. McCarthy, M. (1987). 'Further Additions to the Lichen Flora of the Burren and the Aran Islands (H9),' The Irish Naturalists' Journal, 22(5), 205.
68. McCarthy, M. & Mitchell, M. E. (1985). 'New lichen records from the Burren and the Aran Islands (H9) ,' The Irish Naturalists' Journal, 21(12), 521–523.
69. McGettigan, S. (1972). Some observations on the bark of Corylus avellana and its epiphytic lichen flora in the Burren. B.Sc. Thesis. University College, Galway.
70. McGough, H. N. (1984). A report on the grasslands and closely related vegetation types of the Burren Region of western Ireland. Unpublished report to The Wildlife Service, Dept. of Forest and Fisheries, Dublin.
71. Moles, R & Breen, J. (1991). 'Surveillance of recent vegetation change in the Burren, Co. Clare, Rep. of Ireland,' Global Ecology and Biogeography Letters, 1, 151–156.
72. Moles, R. (1982). 'A Study of Some Bird Communities of the Burren, Co Clare,' The Irish Naturalists' Journal, 20(10), 419–423.
73. Moles, R. (1985). 'Habitats of the Burren,' IQUA Newsletter, 8, 16.
74. Moles, R. (1992). 'Trampling Damage to Vegetation and Soil Cover at Paths within the Burren National Park, Mullach Mor, Co. Clare,' Irish Geography, 25(2), 129–137.
75. Moles, R. (1995). 'Change at the edges of Burren grassland patches,' in Jeffrey, D. W., Jones, M. B. & McAdam, J. H. (eds.) Irish grasslands-their biology and management, Dublin: Royal Irish Academy, 246–252.

76. Moles, R. & Travers, I. (1982). Mullach Mor, Co. Clare, Ireland. Habitat Map.

77. Moles, R., Breen, J. & O'Regan, B. (2005). 'A Pilot Scale Long-Term Experimental Study on the Effects of Grazing and Gap Creation on Burren Grassland Dynamics: Implications for Conservation,' Biology & Environment: Proceedings of the Royal Irish Academy, 105(1), 15–32.

78. Moles, R., Hayes, K., O'Regan, B. & Moles, N. (2003). 'The Impact of Environmental Factors on the Distribution of Plant Species in a Burren Grassland Patch: Implications for Conservation,' Biology & Environment: Proceedings of the Royal Irish Academy, 103(3), 139–145.

79. Molloy, K. & O'Connell, M. (2003). 'Holocene vegetation and land-use dynamics in the karstic environment of Inis Oírr, Aran Islands, western Ireland: pollen analytical evidence evaluated in the light of the archaeological record,' Quaternary International (in press).

80. Morris, M. G. (1967). 'Weevils (Coleoptera, Curculionoidea) and Other Insects Collected in North West Clare, with Special Reference to the Burren Region,' Proceedings of the Royal Irish Academy. Section B: Biological, Geological, & Chemical Science, 65, 349–371.

81. Morris, M. G. (1971). 'Additional records of weevils from the Burren region (Coleoptera: Curculionoidea) ,' The Irish Naturalists' Journal, 17(4), 139–141.

82. Morris, M. G. (1974). 'Auchenorhyncha (Hemiptera) of the Burren, with Special Reference to Species-Associations of the Grasslands,' Proceedings of the Royal Irish Academy. Section B: Biological, Geological, & Chemical Science, 74, 7–30.

83. Morris, M. G. (1974). 'Further records of weevils from the Burren, including two species new to Ireland (Coleoptera: Curculiondea),' The Irish Naturalists' Journal, 18(2), 49–51.

84. Morris, M. G. (1985). 'Bagous brevis Gyllenhal new to Ireland from the Burren, Co. Clare, with a brief review of the Irish Bagoini (Coleoptera: Curculionidae),' The Irish Naturalists' Journal, 21(9), 400–403.

85. Murray, R. (1887). 'Arabis ciliata, R.Br. in Clare,' Journal of Botany, London, 25, 183.

86. Nelson, E. & Stalley, R. A. (1992). 'Medieval Naturalism and the Botanical Carvings at Corcomroe Abbey (Co. Clare),' Gesta, 28(2), 165–174.

87. Nelson, E. C. (1999). Wild plants of the Burren and the Aran Islands. Cork: Collins Press.
88. Nelson, E. C. (2000). An annotated topographical checklist of the flowering plants, conifers, ferns and fern allies of The Burren Region. Dr. E. C. Nelson and The Burren Tourism and Environment Initiative, 2000.
89. Nelson, E. C. & Walsh, W. (1991). The Burren: a companion to the wildflowers of an Irish limestone wilderness. Kilkenny: Boethius Press and The Conservancy of The Burren.
90. Ní Bhriain, B., Gormally, M. & Sheehy Skeffington, M. (2003). 'Changes in Land-use Practices at Two Turloughs, on the East Burren Limestones, Co. Galway, with Reference to Nature Conservation,' Biology & Environment: Proceedings of the Royal Irish Academy, 103(3), 169–176.
91. O'Connell, M. & Jelicic, L. J. (1994). 'Lios Lairthin Mor (LLM II), N. W. Burren: history of vegetation and land use from 3200 BP to the present,' in O'Connell, M. (ed.) Burren, Co. Clare. Irish Association for Quarternary Studies, Field Guide No. 18.
92. O'Donovan (2002). 'Nutrient dynamics of Sesleria-dominated grasslands in the Burren National Park, Co. Clare,' Tearmann, 2, 29–38.
93. O'Mahoney, T. (1860). Notes on botanical excursion in Clare. Proceedings of the Dublin Natural History Society, 1, 30–34.
94. Osborne, B. & Jeffrey, D. W. (2003). 'Understanding the Burren: Introduction and Overview,' Biology and Environment: Proceedings of the Royal Irish Academy, 103B(3), Understanding the Burren, 107–109.
95. Osborne, B., Black, K., Lanigan, G., Perks, M. & Clabby, G. (2003). 'Survival on the Exposed Limestone Pavement in the Burren: Photosynthesis and Water Relations of Three Co–Occurring Plant Species,' Biology & Environment: Proceedings of the Royal Irish Academy, 103(3), 125–137.
96. Parr, S., O'Donovan, G., Ward, S. & Finn, J. A. (2009). 'Vegetation Analysis of Upland Burren Grasslands of Conservation Interest,' Biology & Environment: Proceedings of the Royal Irish Academy, 109(1), 11–33.
97. Platts, E., Bailey, S., McGrath, D. & McGeough, G. (2003). 'A Survey of the Land Winkle Pomatias elegans (Müller, 1774) in the Burren,

Co. Clare,' Biology & Environment: Proceedings of the Royal Irish Academy, 103(3), 197–201.

98. Praeger, R. L. (1903). 'Pinguicula grandiflora in Clare,' Irish Naturalist, 12, 269.

99. Praeger, R. L. (1905). 'Notes on the Botany of Central Clare,' Irish Nature, 14, 188–193.

100. Praeger, R. L. (1932). 'The flora of the turloughs: A preliminary note,' Proceedings of the Royal Irish Academy, 41B, 37–45.

101. Preston, C. D., Stewart, N. F. & Webster, S. D. (1991). 'Records of aquatic plants from Connemara and the Burren,' The Irish Naturalists' Journal, 23(11), 464–467.

102. Proctor, M. C. F. (2010). 'Environmental and vegetational relationships of lakes, fens and turloughs in the Burren,' Biology and Environment: Proceedings of the Royal Irish Academy, 110B, 17–34.

103. Pybus, C., Pybus, M. J. & Ragneborn-Tough, L. (2003). 'Phytoplankton and Charophytes of Lough Bunny, Co. Clare,' Biology & Environment: Proceedings of the Royal Irish Academy, 103(3), 177–185.

104. Roden, C. (1991). 'The Burren Flora,' in O'Connell, J. W. & Korff, A. (eds.) The Book of the Burren, Kinvara: Tir Eolas, 31–43.

105. Rubers, W. V. (1973). Report on the botanical aspects of the Burren, Co. Clare, Ireland. Netherlands Commission for International Nature Protection, Austerlitz.

106. Scannell, M. J. (1972). '"Algal Paper" of Oedogonium sp., Its Occurrence in the Burren, Co. Clare,' The Irish Naturalists' Journal, 17(5), 147–152.

107. Scannell, M. J. (1988). 'Ajuga x hampeana Braun & Vatke in the Burren: B. O'Kelly vindicated,' The Irish Naturalists' Journal, 22(11), 488–490.

108. Shimwell, D. W. (1971). 'Festuco-Bromotea (Br.-Bl. et, R.Tuxen 1943) in the British Isles: The phytogeography and phytosociology of limestone grasslands,' Vegetatio, 23, 1–60.

109. Stewart, S. A. (1889 – 1891) 'Report on the Botany of South Clare and the Shannon,' Proceedings of the Royal Irish Academy, 1, 343–369.

110. Warwick, G. T. (1968). 'Snails from streams and pools in caves at Lisdoonvarna, Co. Clare, Ireland,' Transactions of the Cave Research Group of Great Britain, 10(3), 141.

111. Webb, D. A. (1955). 'Woodland Relics in the Burren,' The Irish Naturalists' Journal, 11(9), 251–252.
112. Webb, D. A. (1963). 'Hymenophyllum wilsonii in the Burren,' The Irish Naturalists' Journal, 14(7), 155–156.
113. Webb, D. A. (1963). 'Noteworthy Plants of the Burren: A Catalogue Raisonné,' Proceedings of the Royal Irish Academy. Section B: Biological, Geological, & Chemical Science, 62, 117–134.
114. Webb, D. A. (1964). Some micro-habitats in the Burren, their micro-environments and vegetation. Proceedings of the Royal Irish Academy. 63B, 291–302.
115. Webb, D. A. (1978). 'The status of Mercurialis perennis in the Burren,' The Irish Naturalists' Journal, 19(6), 200.
116. Webb, D. A. (1982). 'Plant records from Connemara and the Burren,' The Irish Naturalists' Journal, 20(11), 466–471.
117. Webb, D. A. & Scannell, M. J. (1983). Flora of Connemara and The Burren. Royal Dublin Society and Cambridge University Press.
118. Willmot, A. (1979). 'An ecological survey of the ferns of the Burren, Co. Clare, Eire,' Fern Gazette, 12(1), 9–28.

CHAPTER 2

Main Essence of Ecology: Understanding Nature's Interconnected Web

CONTENTS

2.1. INTRODUCTION

Ecology, often described as the study of the relationships between living organisms and their environment, delves into the intricate tapestry of life on Earth. It seeks to unravel the complex interactions that shape ecosystems, populations, and individual species, ultimately revealing the main essence of ecology: the profound interconnectedness of all living things and the environments they inhabit.

At its core, ecology is the science of connections. It is the exploration of how organisms, from the tiniest microbes to the largest mammals, are not isolated entities but integral parts of ecosystems, where they interact with both living and non-living components. To grasp the essence of ecology is to recognize the following key principles:

- Interdependence: The Essence of Ecological Relationships

  The heart of ecology lies in understanding that every organism is interdependent with others within its ecosystem. Species do not exist in isolation but form intricate webs of relationships. For instance, predators rely on prey for sustenance, herbivores shape plant populations, and plants provide oxygen and food for animals. These relationships extend beyond the species level; they encompass entire ecosystems where various organisms, from microorganisms to apex predators, interact.

- Energy Flow: The Vital Thread of Life

  Ecology elucidates the flow of energy through ecosystems. Sunlight, the primary energy source for life on Earth, is captured by photosynthetic organisms like plants, algae, and some bacteria. This energy is then transferred through the food web as organisms consume and are consumed. Energy flow underpins the functioning of ecosystems and shapes their structure, with energy diminishing at each trophic level.

- Nutrient Cycling: Recycling Earth's Elements

  Ecology illuminates the intricate cycles that recycle essential nutrients through ecosystems. Elements like carbon, nitrogen, phosphorus, and water continuously move between the biotic and abiotic realms. Decomposers play a pivotal role in breaking down dead organic matter, and releasing nutrients back into the ecosystem for reuse. Understanding these cycles underscores the importance of maintaining ecological balance.

- Adaptation and Evolution: Nature's Response to Change

  The essence of ecology extends to the evolutionary processes that shape life. Species adapt to their environments over time, and this adaptation can result in the formation of new species. Ecology helps us comprehend how organisms have evolved to occupy specific niches and adapt to various ecological challenges. It showcases the dynamic nature of ecosystems and the constant interplay between species and their surroundings.

- Human Impact: Ecology's Relevance in a Changing World

  In today's world, the main essence of ecology also encompasses humanity's profound influence on the environment. Human activities have led to habitat destruction, pollution, climate change, and biodiversity loss, posing significant challenges to ecosystems and species survival. Ecology is essential for understanding these anthropogenic impacts and devising sustainable solutions to mitigate them.

- Conservation and Sustainability: Preserving Nature's Balance

  Ecology underscores the importance of conservation and sustainability. It offers insights into the consequences of biodiversity loss, the degradation of ecosystems, and the potential collapse of ecological systems. The essence of ecology lies in advocating for responsible environmental stewardship and the protection of Earth's natural resources.

- Holistic Perspective: The Essence of Systems Thinking

  Ecology encourages a holistic perspective. It recognizes that all components of an ecosystem are interconnected, and changes in one part can ripple throughout the system. This systems thinking is crucial for understanding the complexity of ecological processes and for making informed decisions about environmental management and policy.

So, the main essence of ecology lies in recognizing the profound interconnectedness of all living organisms and their environments. It teaches us that nature is a vast web of relationships, where every species, every individual, and every element plays a role in maintaining the delicate balance of life on Earth. As stewards of this planet, understanding and respecting this essence is vital for preserving the integrity and health of our ecosystems and ensuring a sustainable future for all living beings. Ecology provides us with the knowledge and tools to achieve this noble goal.

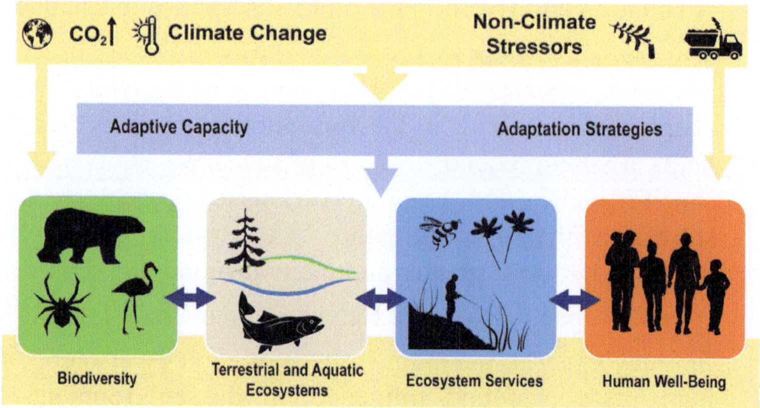

Figure 4. Essence of Ecology: Nature's Interconnected Web

(Source: Pielke, R. A. Sr., Peters, D. P. C., & Niyogi, D. Ecology and Climate of the Earth—The Same Biogeophysical System. Climate. 2022; 10(2):25. https://doi.org/10.3390/cli10020025, Creative Commons License)

2.2. NATURE

Figure 5. Nature

(Source: By Photo by David Iliff, CC BY-SA 3.0, https://commons.wikimedia.org/w/index.php?curid=43330268, Creative Commons License)

Ecology, as a scientific discipline, revolves around the study of nature and the relationships between living organisms and their environment. Therefore, it is intrinsic to the essence of ecology that "nature" is not only a fundamental concept but also the very essence from which the discipline derives its significance and purpose. In this detailed analysis, we explore the profound connection between ecology and the concept of nature, highlighting the pivotal role of nature in ecological understanding and research.

Nature as the Study Subject:

- Natural Systems: Ecology primarily focuses on natural systems, which encompass all living organisms, ecosystems, landscapes, and biophysical processes. Nature, in its pristine form, serves as the subject of ecological inquiry. Ecologists seek to unravel the mysteries of these systems and understand how they function.
- Biodiversity: The immense diversity of life forms found in nature is a central theme in ecology. Ecologists study the intricate patterns of biodiversity, from the microscopic to the macroscopic, and explore the factors that influence species richness and distribution.

Nature as a Source of Knowledge:

- Empirical Observations: Ecological research often involves fieldwork and direct observation of natural phenomena. The data collected from nature forms the empirical basis for ecological theories and models. Ecologists draw on their observations to formulate hypotheses and test them rigorously.
- Long-Term Studies: Long-term ecological studies conducted in natural environments provide invaluable insights into ecological processes, including succession, population dynamics, and adaptation to changing conditions. These studies are essential for understanding the resilience and stability of ecosystems.

Nature as an Inspirational Model:

- Bioinspiration: Nature serves as an inspiration for human innovation and technology. Ecologists and engineers collaborate to develop sustainable solutions by emulating nature's efficiency and design principles. Examples include biomimicry in architecture, transportation, and materials science.
- Ecosystem Services: Nature provides essential ecosystem services such as clean air, clean water, pollination, and soil

fertility. Ecological research quantifies the value of these services, emphasizing the interdependence of human well-being and nature.

Nature as an Ethical Imperative:

- Conservation and Preservation: Ecology plays a critical role in advocating for the conservation and preservation of natural habitats and biodiversity. The ethical imperative of protecting nature stems from an understanding of its intrinsic value and the interconnectedness of all life forms.
- Sustainability: Ecological principles underpin sustainability practices, emphasizing the need to maintain the health and resilience of natural systems. The concept of sustainable development recognizes that human well-being is intertwined with the health of the environment.

Nature as a Complex System:

- Complexity and Interconnectedness: Nature is inherently complex, characterized by intricate networks of ecological interactions. Understanding nature requires acknowledging the interconnectedness of species, trophic levels, and abiotic factors. This complexity challenges ecologists to adopt systems thinking.
- Emergent Properties: Nature exhibits emergent properties, where the behavior of ecological systems cannot be solely explained by studying individual components. These emergent properties, such as self-organization and feedback loops, are fundamental to ecology's understanding of nature.

Nature as a Context for Change:

- Environmental Change: Ecology examines how natural systems respond to environmental changes, including disturbances, climate fluctuations, and human impacts. The study of nature's resilience and adaptability is crucial for addressing global challenges such as climate change.
- Restoration and Rehabilitation: Ecological restoration efforts aim to repair and rejuvenate ecosystems that have been damaged by human activities. These initiatives rely on a deep understanding of nature's processes and the importance of restoring natural balance.
- In essence, nature is not only the subject of ecology but also its

guiding principle, moral compass, and wellspring of knowledge. The discipline of ecology thrives on the recognition that nature is a complex, interconnected, and ever-changing tapestry of life that warrants careful study, preservation, and responsible management. Ecologists dedicate themselves to unraveling the mysteries of nature, seeking solutions to environmental challenges, and advocating for the sustainable coexistence of humans and the natural world. In doing so, they honor the main essence of ecology—nature itself.

2.3. NATURAL ENVIRONMENT

Figure 6. Natural Environment

(Source: By Ryan Gsell – originally posted to Flickr as DSC01688, CC BY 2.0, https://commons.wikimedia.org/w/index.php?curid=7309354, Creative Commons License)

Ecology, as a field of science, is fundamentally rooted in the study of the natural environment and its interactions with living organisms. It is through the lens of the natural environment that ecology gains its significance, purpose, and relevance. In this detailed analysis, we explore the central role of the natural environment in ecology, highlighting how it forms the bedrock of ecological understanding and research.

The Natural Environment as the Subject of Study:

- Ecosystems and Habitats: Ecology is intrinsically concerned with ecosystems and habitats found in the natural environment. These diverse ecosystems, including forests, oceans, grasslands, and wetlands, serve as the laboratories where ecologists conduct their investigations.

- Biotic and Abiotic Components: Ecologists examine the intricate relationships between the biotic (living) and abiotic (non-living) components of the natural environment. These interactions encompass a wide array of ecological processes, from nutrient cycling to predation and competition.

The Natural Environment as a Source of Data and Knowledge:

- Field Research: Ecological research often relies on fieldwork, where scientists directly observe and collect data from the natural environment. These field observations and experiments provide the empirical foundation for ecological theories and models.
- Biodiversity and Species Interactions: The natural environment is a wellspring of biodiversity, housing countless species. Ecological studies conducted in the field reveal the richness of species interactions, from symbiosis to trophic relationships.

The Natural Environment as an Inspiration and Model:

- Bioinspiration: Nature's designs, mechanisms, and strategies have inspired technological and engineering innovations. Biomimicry, a practice that emulates nature's solutions to complex problems, has resulted in sustainable designs and products.
- Ecosystem Services: The natural environment provides essential ecosystem services, including clean air, water purification, pollination, and climate regulation. Recognizing the value of these services underscores the deep interconnection between human well-being and the environment.

The Natural Environment as an Ethical Imperative:

- Conservation and Preservation: Ecology plays a critical role in advocating for the conservation and preservation of natural habitats and species. It stems from an ethical imperative to protect the intrinsic value of the natural environment and maintain the integrity of ecosystems.
- Sustainable Practices: Ecology underpins sustainability practices by emphasizing the need to maintain the health and resilience of natural systems. It promotes the responsible use of natural

resources to ensure the well-being of present and future generations.

The Natural Environment as a Complex and Dynamic System:

- Complexity and Dynamics: The natural environment is characterized by its complexity and dynamism. Ecological systems exhibit intricate patterns of interaction and change over time. Understanding these complexities is central to ecology's mission.
- Resilience and Adaptation: Ecology explores how natural systems respond to disturbances and environmental changes. The study of resilience and adaptation in the natural environment is crucial for addressing pressing challenges such as climate change and habitat degradation.

The Natural Environment as the Context for Change:

- Environmental Challenges: Ecology investigates how natural systems are affected by environmental challenges, including human-induced changes such as pollution, deforestation, and habitat loss. It seeks solutions to mitigate these impacts and restore ecological balance.
- Environmental Education and Advocacy: A deep understanding of the natural environment fosters environmental education and advocacy. Ecologists and conservationists strive to raise awareness about the importance of nature and the need for its protection.

Thus, the natural environment serves as the cornerstone of ecology. It is not merely a subject of study but the very essence from which ecological knowledge, principles, and ethics emerge. Ecology acknowledges that the natural environment is a complex, interconnected, and dynamic system that warrants careful examination, conservation, and responsible stewardship. Ecologists, as custodians of this discipline, recognize the profound interdependence between the health of the natural environment and the well-being of all living organisms, including humans. They dedicate their efforts to unravelling the mysteries of nature, preserving its integrity, and advocating for its sustainable management—a testament to the central and enduring role of the natural environment in ecology.

2.4. ECOSYSTEM

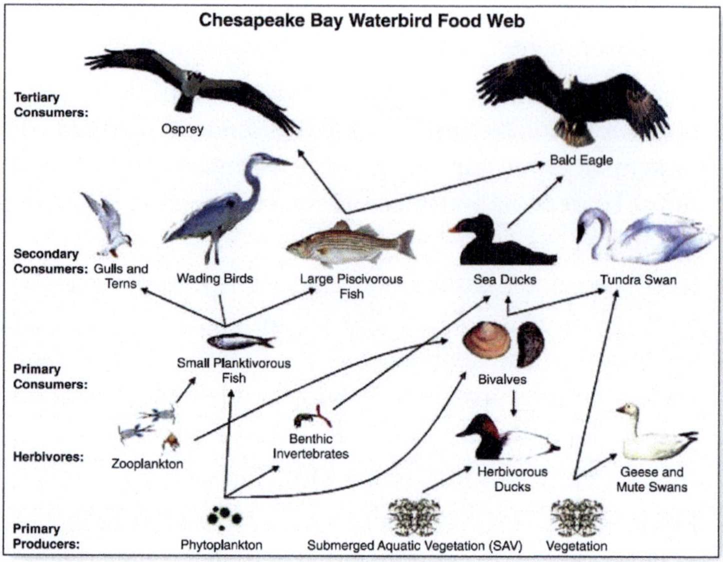

Figure 7. Ecosystem

(Source: By Matthew C. Perry – US Geological Survey (USGS)."Chapter 14: Changes in Food and Habitats of Waterbirds." Figure 14.1. Synthesis of U.S. Geological Survey Science for the Chesapeake Bay Ecosystem and Implications for Environmental Management. USGS Circular 1316., Public Domain, https://commons.wikimedia.org/w/index.php?curid=10491372, Creative Commons License)

Ecology, as a scientific discipline, finds its primary essence in the study of ecosystems. Ecosystems serve as the fundamental units of ecological investigation, understanding, and appreciation. In this detailed analysis, we explore the central role of the ecosystem in ecology, highlighting how it encapsulates the core principles and focus of the discipline.

Ecosystems as the Primary Units of Study:

- Definition of Ecosystems: Ecosystems are defined as complex, interactive systems comprising living organisms (biotic components) and their physical environment (abiotic components) within a specific geographical area. These areas can range from a tiny pond to a vast rainforest.
- Ecological Levels: Ecosystems encapsulate multiple ecological levels, from individuals and populations to communities and

landscapes. By examining these levels within ecosystems, ecologists gain a comprehensive understanding of ecological processes.

Biotic and Abiotic Interactions:

- Dynamic Interplay: Ecosystems are characterized by dynamic interactions between living organisms and their abiotic surroundings. These interactions encompass factors such as nutrient cycling, energy flow, predation, competition, and adaptation.
- Keystone Species: Ecosystems often feature keystone species that exert disproportionate influence on their structure and function. These species, whether large predators or tiny microbes, play critical roles in maintaining ecological balance.

Ecosystem Processes and Functions:

- Energy Flow: Ecosystems demonstrate the flow of energy through trophic levels, starting with primary producers (e.g., plants) that capture sunlight through photosynthesis. This energy is subsequently transferred to herbivores, predators, and decomposers.
- Nutrient Cycling: Ecosystems exhibit the cycling of essential nutrients (e.g., carbon, nitrogen, phosphorus) through biological, geological, and chemical processes. These nutrient cycles sustain life within the ecosystem.

Biodiversity and Species Interactions:

- Species Diversity: Ecosystems showcase a rich array of species, each with its ecological niche. Biodiversity within ecosystems is a central focus of ecological research, encompassing species richness, evenness, and ecological roles.
- Interactions: Ecosystems harbor diverse ecological interactions, including competition, predation, mutualism, and parasitism. These interactions shape species composition and population dynamics.

Ecosystem Services and Human Well-being:

- Human Dependence: Ecosystems provide crucial ecosystem services, including clean water, air purification, pollination, and climate regulation. These services are essential for human well-being, agriculture, and industry.

- Economic and Cultural Value: Ecosystems hold economic and cultural significance for human societies. They support livelihoods, provide recreational opportunities, and inspire cultural practices and traditions.

Ecosystem Health and Resilience:

- Health Indicators: Ecologists assess ecosystem health through indicators such as species diversity, habitat integrity, and water quality. Healthy ecosystems are resilient and better equipped to withstand environmental changes.
- Restoration and Conservation: Understanding ecosystem dynamics is critical for conservation and restoration efforts. Ecologists play a pivotal role in identifying strategies to conserve and rehabilitate damaged ecosystems.

Ecosystem Sustainability:

- Sustainable Management: Ecosystem-based management emphasizes sustainable practices that balance human needs with the preservation of ecosystem integrity. It recognizes that ecosystems are finite and need protection to ensure long-term sustainability.
- Complexity and Adaptation: Ecosystems' inherent complexity requires adaptive management strategies that account for natural variability and change. Ecosystem-based approaches adapt to evolving ecological conditions.

In essence, ecosystems epitomize the core of ecology. They embody the intricate and dynamic relationships between living organisms and their environment. Ecology, as a discipline, centres on the study of these ecosystems to unravel the mysteries of life, understand the consequences of environmental changes, and advocate for responsible stewardship of Earth's natural systems. Ecosystems encapsulate the essence of ecological research, conservation, and sustainability—a testament to their paramount importancein 'he realm of ecology.

2.5. BIOME

Ecology, a science deeply concerned with understanding the relationships between living organisms and their environment, finds one of its most significant focal points in the concept of biomes. Biomes represent large-scale ecological regions characterized by distinct climates, vegetation, and wildlife. In this detailed analysis, we explore the central role of biomes in ecology, emphasizing how they encapsulate core ecological principles and serve as vital units for ecological understanding and research.

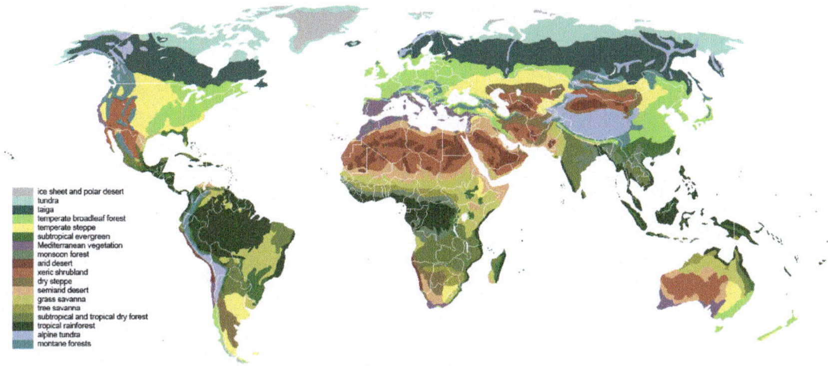

Figure 8. Biome

(Source: By Ville Koistinen (user Vzb83) – the blank world map in Commons and WSOY Iso karttakirja for the information, CC BY-SA 3.0, https://commons.wikimedia.org/w/index.php?curid=1700408, Creative Commons License)

Defining Biomes:

- Distinctive Regions: Biomes are defined as distinctive regions of the Earth's surface with specific climate patterns and characterized by particular types of vegetation, which in turn influence the organisms that inhabit these areas.
- Terrestrial and Aquatic Biomes: While terrestrial biomes relate to land-based ecosystems, aquatic biomes encompass various aquatic environments such as freshwater lakes, rivers, and marine ecosystems. Each biome exhibits unique features and ecological dynamics.

Biodiversity and Adaptation:

- Species Diversity: Biomes showcase remarkable biodiversity, with various species adapted to the specific conditions of each biome. These adaptations often include physiological, behavioral, and morphological traits that enable organisms to thrive in their respective biomes.
- Endemism: Many biomes harbor endemic species—organisms found exclusively within a particular biome. These endemic species have evolved to fill specific ecological niches within their biome.

Climate and Vegetation:

- Climate Influence: Climate is a defining factor for biomes, shaping their characteristics. Temperature, precipitation, and seasonality profoundly impact the vegetation and wildlife in each biome.
- Distinct Vegetation: Each biome is associated with a particular type of vegetation, such as tropical rainforests, temperate deciduous forests, deserts, or tundras. These vegetation types, in turn, influence the availability of resources and habitats for organisms.

Ecological Interactions:

- Species Relationships: The interactions among species within biomes include predation, competition, mutualism, and parasitism. These relationships have a profound impact on the structure and dynamics of biome ecosystems.
- Food Webs and Trophic Levels: Biomes host complex food webs with multiple trophic levels, from producers (plants) to primary consumers (herbivores) and secondary consumers (predators). These interactions dictate energy flow within the biome.

Human Impact and Conservation:

- Human Footprint: Human activities, such as deforestation, urbanization, and resource extraction, have significant impacts on biomes. Human-induced changes in biomes can result in biodiversity loss and ecological disruption.
- Conservation Efforts: Conservation efforts often target biomes as focal points for protection and restoration. Conservationists work to safeguard the unique biodiversity and ecological functions of these regions.

Biome Shifts and Climate Change:

- Climate Change Effects: Climate change can lead to shifts in biome boundaries and characteristics as temperature and precipitation patterns alter. These shifts have consequences for species distributions and ecosystem dynamics.
- Resilience and Adaptation: Understanding biome resilience and the adaptive capacity of species within them is crucial for addressing the challenges posed by climate change.

Hence, biomes represent the essence of ecology by encapsulating the diversity and complexity of Earth's ecosystems on a grand scale. They serve as ecological units that bring together climate, vegetation, wildlife, and intricate ecological interactions. Biomes provide ecologists with a framework for understanding how the environment shapes life on our planet and how organisms, in turn, influence the structure and function of these ecosystems. The study of biomes is essential for ecological research, conservation, and our broader understanding of the interconnectedness of Earth's natural systems.

2.5.1. Community (Ecology)

In the field of ecology, a "community" refers to an essential concept that plays a central role in understanding how ecosystems function and how species interact within them. A community represents a group of populations of different species that coexist in the same geographic area and interact with one another. Here, we provide a detailed analysis of the concept of a community in ecology:

Community Composition:

- Species Diversity: A community is characterized by its species composition, including the number of species present and their relative abundance. Species diversity is a key aspect of community ecology, and it can vary widely from one community to another.
- Species Interactions: Within a community, species interact in various ways. These interactions can be classified as competition (for limited resources), predation (one species consuming another), mutualism (mutually beneficial interactions), parasitism (one species benefiting at the expense of another), and more.

Community Structure:

- Trophic Structure: Community structure often involves a trophic hierarchy, which describes the feeding relationships among species. Producers (plants) are at the base of the food web, followed by herbivores, carnivores, and decomposers.
- Keystone Species: Some species within a community have a disproportionately large impact on its structure and function. These keystone species play vital roles in maintaining the diversity and stability of the community. Their removal can trigger significant ecological changes.

Species Interactions:

- Competition: Competition for limited resources (e.g., food, habitat, and mates) is a fundamental ecological interaction within communities. It can lead to resource partitioning, where species evolve to use resources differently to reduce competition.

- Predation: Predation and herbivory have a significant influence on community structure. Predators control prey populations, which can have cascading effects on other species within the community.
- Mutualism: Mutualistic interactions, where species benefit from their association with one another, are common in communities. Examples include pollination, seed dispersal, and nitrogen-fixing bacteria in plant roots.

Biotic and Abiotic Factors:

- Biotic Interactions: Interactions between species (biotic factors) play a critical role in shaping community structure. These interactions can result in coevolution, where species evolve in response to one another.
- Abiotic Factors: Abiotic factors, such as temperature, precipitation, and soil composition, also influence community composition and distribution. These factors determine which species can thrive in a particular environment.

Succession and Disturbance:

- Ecological Succession: Communities can change over time through a process known as ecological succession. This can be primary succession (starting from bare rock or soil) or secondary succession (following disturbances like fires or logging). Succession influences the composition and diversity of a community.
- Disturbance: Disturbances, whether natural (e.g., wildfires, storms) or human-induced (e.g., deforestation), can disrupt community structure. However, some communities are adapted to frequent disturbances and have a high degree of resilience.

Ecological Resilience:

- Community Resilience: The ability of a community to resist change or bounce back after a disturbance is referred to as community resilience. Resilient communities can maintain their structure and function even in the face of environmental challenges.

Conservation and Restoration:

- Conservation Importance: Understanding the dynamics of communities is crucial for biodiversity conservation. Conservation efforts often focus on preserving entire communities to protect the species they contain.
- Restoration Ecology: Restoring damaged or degraded communities is a central goal of restoration ecology. This involves reintroducing native species and rehabilitating ecosystems to promote ecological balance.

Therefore, the concept of a community in ecology represents a fundamental unit of study for understanding how species interact within ecosystems. It encompasses the diverse array of species, their interactions, and the structure and dynamics of the biological assemblage within a defined geographic area. Communities play a pivotal role in ecological research, conservation, and our broader understanding of how ecosystems function in the natural world.

2.5.2. Biocoenosis, Species, Population, Organism: An Ecological Hierarchy

In ecology, the concepts of biocoenosis, species, population, and organism represent a hierarchical framework for understanding the structure and interactions within ecological systems. Let's explore these concepts and their relationships:

Organism:

- An organism is the most basic unit in the hierarchy of ecological organization. It refers to an individual living entity, which can be a plant, an animal, a microbe, or any other form of life.
- Organisms exhibit specific traits, behaviors, and physiological characteristics that define their species and determine their interactions with the environment.

Population:

- A population consists of a group of individuals of the same species living in the same geographic area at the same time.
- Populations represent the next level of ecological organization above the individual organism. They are characterized by factors such as birth rates, death rates, immigration, and emigration.

- Ecologists study populations to understand how they grow, fluctuate, and respond to environmental changes.

Species:

- A species is a group of organisms that are capable of interbreeding and producing fertile offspring under natural conditions.
- Species are a fundamental unit of ecological classification and represent a level of organization that encompasses populations of individuals with similar genetic and ecological characteristics.
- Biodiversity, a key concept in ecology, is often measured by the number of species within a given ecosystem or geographical region.
- **Biocoenosis:**
- Biocoenosis, also known as a biological community, represents the highest level in this ecological hierarchy.
- It comprises all the populations of different species that inhabit a specific area and interact with each other.
- In biocoenosis, species populations are interconnected through various ecological relationships, including predation, competition, mutualism, and more.
- The study of biocoenosis is central to understanding how species interact within ecosystems, influence one another, and collectively shape the ecological dynamics of a given area.

Interactions and Significance:

- Organisms make up populations, populations constitute species, and species collectively form biocoenosis or biological communities. This hierarchical framework is essential for ecologists to analyze and comprehend the complexity of ecosystems.
- Interactions within and between these ecological levels drive the functioning and dynamics of ecosystems. For example, the growth and stability of a population depend on the individual organisms that comprise it, and the structure of a biological community is influenced by the composition and interactions of the species within it.
- The study of these ecological levels is crucial for various ecological disciplines, including population ecology, community

ecology, and conservation biology. It allows scientists to investigate questions related to species diversity, population dynamics, ecosystem structure, and the effects of environmental change on biological systems.

In summary, the concepts of biocoenosis, species, population, and organism represent a hierarchical structure that forms the basis of ecological understanding. They provide a framework for exploring the interrelationships and dynamics within ecological systems, from individual organisms to complex biological communities. This hierarchical approach is vital for studying the intricate web of life on Earth and for addressing ecological questions and challenges in a systematic manner.

2.6. BIODIVERSITY

Figure 9. Biodiversity

(Source: https://commons.wikimedia.org/w/index.php?curid=20426804, Creative Commons License)

Biodiversity, a contraction of "biological diversity," is a concept that encapsulates the variety of life forms on our planet. It refers to the incredible diversity of living organisms, including plants, animals, fungi, and microorganisms, the ecological complexes they form, and the genetic diversity found within species. Biodiversity is a fundamental pillar of life on Earth and plays a pivotal role in the health and stability of ecosystems. Here, we delve into the multifaceted world of biodiversity and its critical importance:

Levels of Biodiversity:

- Genetic Diversity: Genetic diversity refers to the variation in genes within a species. It is the raw material for evolution, enabling species to adapt to changing environmental conditions. Genetic diversity is crucial for the long-term survival of species.
- Species Diversity: Species diversity represents the variety of species within a given area or ecosystem. It encompasses the number of species present, their relative abundance, and the distribution patterns of different species.
- Ecosystem Diversity: Ecosystem diversity pertains to the variety of ecosystems, such as forests, grasslands, wetlands, coral reefs, and deserts, found on Earth. Each ecosystem has unique physical and biological characteristics.

Importance of Biodiversity:

- Ecosystem Services: Biodiversity is the foundation of ecosystem services, which include provisioning services (e.g., food and medicine), regulating services (e.g., climate regulation and pest control), supporting services (e.g., nutrient cycling), and cultural services (e.g., recreation and aesthetics).
- Stability and Resilience: Diverse ecosystems tend to be more stable and resilient in the face of environmental changes. Biodiversity provides a buffer against disturbances and helps ecosystems recover after disruptions.
- Economic Value: Biodiversity has economic significance. Many industries, such as agriculture and pharmaceuticals, rely on diverse genetic resources for crop breeding and drug discovery.
- Cultural and Aesthetic Value: Biodiversity enriches human cultures, traditions, and aesthetics. It inspires art, literature, and spirituality, and it provides recreational opportunities like wildlife viewing and ecotourism.

Threats to Biodiversity:

- Habitat Destruction: The primary driver of biodiversity loss is habitat destruction, including deforestation, urbanization, and conversion of natural landscapes into agriculture.
- Pollution: Pollution from various sources, including industrial, agricultural, and urban runoff, can harm ecosystems and species.
- Overexploitation: Overharvesting of natural resources, such as

overfishing and illegal wildlife trade, poses a significant threat to biodiversity.

- Climate Change: Climate change alters habitats and disrupts species distributions. It can lead to the loss of specialized habitats and exacerbate biodiversity decline.

Conservation and Preservation:

- Protected Areas: Establishing and maintaining protected areas, such as national parks and wildlife reserves, is a crucial conservation strategy to safeguard biodiversity.
- Habitat Restoration: Efforts to restore degraded habitats, such as reforestation and wetland restoration, help rejuvenate ecosystems and support biodiversity recovery.
- Sustainable Practices: Promoting sustainable practices in agriculture, fisheries, and forestry can reduce the negative impacts on biodiversity while ensuring human well-being.
- Education and Awareness: Raising public awareness about the value of biodiversity and the importance of conservation is essential for fostering a sense of responsibility for preserving life on Earth.

Global Initiatives:

- Convention on Biological Diversity (CBD): The CBD is an international treaty aimed at conserving biodiversity, promoting sustainable use of biological resources, and ensuring equitable sharing of benefits from genetic resources.
- Sustainable Development Goals (SDGs): Biodiversity conservation is integral to several of the United Nations' SDGs, particularly Goal 15, which focuses on life on land, and Goal 14, which focuses on life below water.

Biodiversity, often referred to as Earth's "life insurance policy," is under threat due to human activities. It is essential to recognize the value of biodiversity, not only for ecological stability but also for human well-being. Preserving and restoring biodiversity is a global imperative, and concerted efforts are needed to ensure that the rich tapestry of life on Earth continues to thrive for generations to come.

2.6.1. Food Web

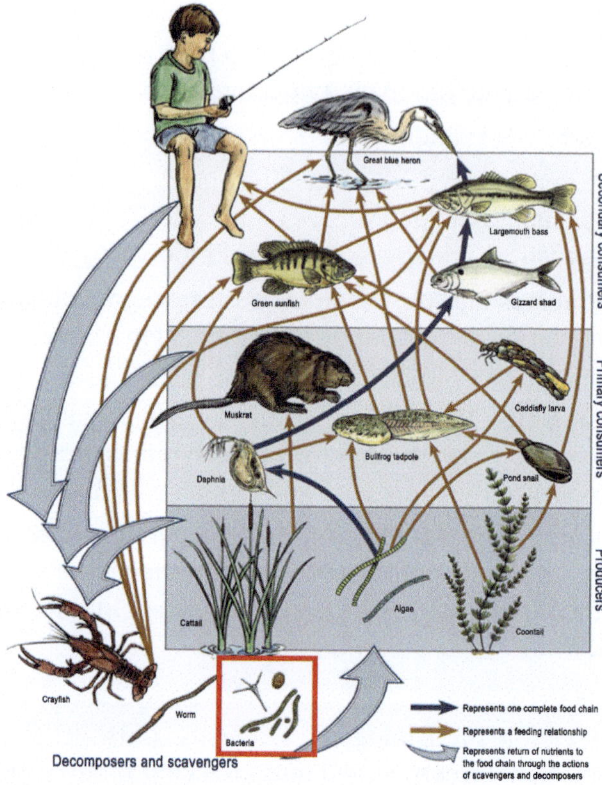

Figure 10. Food Chain & Web

(Source: By Kestin Schulz, Mariya W. Smit, Lydie Herfort and Holly M. Simon. The image (amended) was provided courtesy of the Missouri Department of Conservation. – [1] doi:10.3389/frym.2018.00004, CC BY-SA 4.0, https://commons.wikimedia.org/w/index.php?curid=97329917, Creative Commons License)

A food web is a graphical representation of the intricate network of feeding relationships within an ecosystem. It depicts how various species, including plants, animals, and microorganisms, interact with one another by consuming and being consumed. Food webs are fundamental to ecology as they illustrate the flow of energy and the transfer of nutrients among different organisms within an ecosystem. Here, we delve into the concept of a food web and its significance in ecological systems:

Components of a Food Web:

- Producers: At the base of every food web are producers, typically green plants or photosynthetic microorganisms. They convert sunlight into chemical energy through photosynthesis, forming the foundational energy source for the entire ecosystem.
- Consumers: Consumers are organisms that feed on other organisms. They can be grouped into several categories based on their feeding habits:
- Herbivores: Herbivores are primary consumers that feed exclusively on plants (e.g., grazing mammals).
- Carnivores: Carnivores are secondary and tertiary consumers that primarily eat other animals (e.g., predators like lions).
- Omnivores: Omnivores have a varied diet, consuming both plants and animals (e.g., humans and bears).
- Detritivores and Decomposers: These organisms break down dead organic matter (detritus) into simpler substances. Detritivores, such as earthworms, feed on partially decomposed material, while decomposers like bacteria and fungi complete the decomposition process.

Trophic Levels:

Trophic levels represent the hierarchical feeding positions within a food web. They typically include:

- Producers (Trophic Level 1): These are the primary energy converters, occupying the lowest trophic level.
- Primary Consumers (Trophic Level 2): Herbivores that feed directly on producers.
- Secondary Consumers (Trophic Level 3): Carnivores that feed on herbivores.
- Tertiary Consumers (Trophic Level 4): Predators that feed on secondary consumers.
- Quaternary Consumers (Trophic Level 5): In some cases, additional trophic levels may exist, with predators consuming tertiary consumers.

Complexity and Interconnectivity:

Food webs are highly complex, reflecting the multitude of species interactions within an ecosystem. Species often have multiple food sources and predators, resulting in intricate interconnectedness.

Keystone species, which have a disproportionate influence on community structure, can exert substantial control over the dynamics of a food web. Their presence or absence can trigger cascading effects throughout the ecosystem.

Energy Flow and Nutrient Cycling:

Energy flows through a food web in a unidirectional manner, from producers to consumers and ultimately to decomposers. As energy is transferred from one trophic level to the next, some is lost as heat, resulting in a pyramid-shaped energy transfer.

Nutrient cycling is another critical aspect of food webs. When organisms die and are decomposed by detritivores and decomposers, nutrients are released back into the ecosystem, sustaining primary production by plants and algae.

Significance in Ecology:

Food webs are central to understanding ecosystem dynamics, including population regulation, energy transfer, and species interactions.

They provide insights into the stability and resilience of ecosystems, as changes in one species or trophic level can have cascading effects throughout the web.

Food webs are essential tools for ecological research and conservation efforts. They help scientists assess the impacts of disturbances, such as habitat loss or invasive species, on ecosystem structure and function.

So, food webs are intricate representations of the feeding relationships that govern ecosystems. They illustrate the complex interactions among species, energy flow through trophic levels, and the recycling of nutrients within ecological systems. Understanding and analyzing food webs are essential for ecologists seeking to unravel the mysteries of ecological processes, biodiversity conservation, and the delicate balance of life in Earth's diverse ecosystems.

2.7. REFERENCES

1. Armsworth, P. R., Armsworth, A. N., Compton, N., Cottle, P., Davies, I., Emmett, B. A., Fandrich, V., Foote, M., Gaston, K. J., Gardiner, P., Hess, T., Hopkins, J., Horsley, N., Leaver, N., Maynard, T., & Shannon, D. (2010). The ecological research needs of business. Journal of Applied Ecology, 47, 235–243.
2. Begon, M., Townsend, C. R., & Harper, J. L. (2006). Ecology, From Individuals to Ecosystems. Malden: Blackwell.
3. Crane, A., & Matten, D. (2007). Business Ethics (2nd ed.). New York: Oxford University Press.
4. Crane, A., McWilliams, A., Matten, D., Moon, J., & Siegel, D. S. (2008). The Oxford Handbook of Corporate Social Responsibility. Oxford/New York: Oxford University.
5. Global Footprint Network (2010). Ecological Footprint Atlas 2010. www.footprintnetwork.org. (accessed on 29 September 2023)
6. Goh-Jeevan, J., & Loh, B. (2011). A new shift in CSR. Palm oil industry's role in biodiversity protection, conservation and natural resource management. Journal of Oil Palm & The Environment, 2, 63–71.
7. Loeffler, M., & Roeder, I. (2002). Tissue stem cells: Definition, plasticity, heterogeneity, self-organization and models – A conceptual approach. Cell Tissue Organs, 171, 8–26.
8. Odum, E. P., & Barrett, G. W. (2005). Fundamentals of Ecology (5th ed.). Belmont: Thomson Brooks/Cole.
9. Porter, M. E., & Kramer, M. R. (2006). Strategy and society: The link between competitive advantage and corporate social responsibility. Harvard Business Review, 78–92.
10. Shen, C., & Chang, Y. (2009). Ambition versus conscience, does corporate social responsibility pay off? the application of matching methods. Journal of Business Ethics, 88(1), 133–153.
11. Visser, W. (2011). The Age of Responsibility: CSR 2.0 and the New DNA of Business. Chichester: Wiley.

CHAPTER 3

History of Ecology

CONTENTS

3.1. INTRODUCTION

Ecology, as a scientific discipline, has a rich history that has evolved over centuries. The study of ecological principles and the development of ecological thought have been influenced by various historical events, thinkers, and discoveries. Here are some research headings for the history of ecology, along with detailed analysis for each:

Early Roots of Ecology

This section explores the origins of ecological thought in ancient civilizations, including Indigenous knowledge systems, early observations of natural patterns, and the writings of early philosophers like Aristotle, who made significant contributions to the understanding of organisms and their environments.

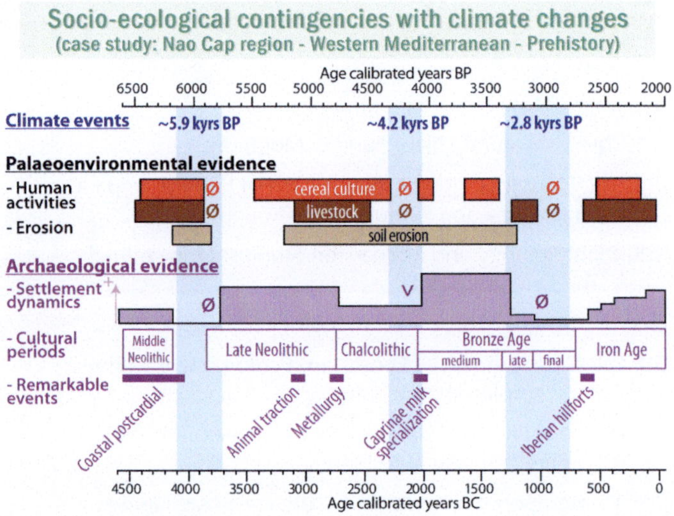

Figure 11. History of Ecology

(Source: Brisset, E., Revelles, J., Expósito, I., Bernabeu Aubán, J., & Burjachs, F. (2020). Socio-Ecological Contingencies with Climate Changes over the Prehistory in the Mediterranean Iberia. Quaternary. 3(3), 19. https://doi.org/10.3390/quat3030019, Creative Commons License)

Rise of Natural History and Exploration

During the Age of Exploration, naturalists and explorers began documenting the flora and fauna of newly discovered lands. This period laid the foundation for modern ecology by cataloging species diversity and observing ecological relationships.

The Darwinian Revolution

Charles Darwin's theory of evolution by natural selection, presented in "On the Origin of Species," revolutionized biology and had profound implications for ecology. This heading explores how Darwin's ideas shaped the study of ecosystems and the interplay between species.

Plant Ecology and Succession Studies

The late 19th and early 20th centuries witnessed significant advancements in plant ecology and studies on ecological succession. Pioneers like Frederic Clements and Henry Cowles made substantial contributions to our understanding of community dynamics.

The Birth of Conservation Ecology

The late 19th century also saw the emergence of conservation movements in response to habitat destruction and species extinction. This heading examines the role of early conservationists like John Muir and the establishment of the world's first national parks.

The Development of Ecosystem Ecology

The mid-20th century marked the rise of ecosystem ecology, which focused on studying energy flow and nutrient cycling within ecosystems. Pioneering ecologists like Raymond Lindeman and Eugene Odum played pivotal roles in this development.

Modern Environmentalism and Conservation Biology

The latter half of the 20th century saw a surge in environmental awareness, leading to the modern environmental movement. This heading explores the role of figures like Rachel Carson, whose book "Silent Spring" sparked global concern about pesticide use, and the emergence of conservation biology as a discipline.

Contemporary Ecological Challenges

This research heading examines the major ecological challenges of our time, including climate change, habitat loss, biodiversity decline, and the application of ecological principles to address these issues.

Role of Technology and Big Data in Ecology

With the advent of advanced technology and data analysis, contemporary ecologists are able to collect and process vast amounts of ecological data. This heading explores the impact of technology on ecological research and conservation efforts.

Interdisciplinary Approaches in Modern Ecology

Ecological research has increasingly become interdisciplinary, incorporating fields like genetics, economics, and social sciences. This heading delves into the evolution of ecological research methodologies and approaches.

Future Directions in Ecology

This research heading speculates on the future of ecology, including emerging trends such as urban ecology, synthetic biology, and the application of ecological principles to address global challenges.

By delving into the history of ecology through these research headings, scholars can gain a comprehensive understanding of how ecological thought has developed over time, how it has been shaped by historical events and influential figures, and how it continues to evolve in response to contemporary ecological challenges.

3.2. EARLY ROOTS OF ECOLOGY

The history of ecology can be traced back to ancient civilizations and early philosophical ideas that laid the foundation for our contemporary understanding of the natural world. "Early Roots of Ecology" explores the origins of ecological thought and the intellectual journey that led to the emergence of ecology as a distinct scientific discipline.

Knowledge

Indigenous cultures around the world have long maintained a deep connection with their environments. Their knowledge of ecosystems, sustainable practices, and the interdependence of species has played a vital role in shaping early ecological insights.

Indigenous knowledge systems, such as the traditional ecological knowledge of Native American tribes, the wisdom of Aboriginal Australians, and the practices of Inuit communities, reflect profound ecological understanding.

Ancient Philosophical Foundations

Early Greek philosophers like Aristotle (384–322 BCE) made significant contributions to ecological thought. Aristotle's works, such as "Historia Animalium" and "Parts of Animals," laid the groundwork for the classification and study of organisms in their natural habitats.

Aristotle introduced concepts like "ecosystem" (although the term wasn't used) and recognized the interconnectedness of species within a community. He also discussed the roles of organisms in maintaining ecosystem balance.

Historical Observations and Manuscripts

Throughout history, naturalists and scholars documented their observations of nature. Historical manuscripts, including travelogues and botanical writings, contain valuable insights into early ecological observations.

The works of explorers like Charles Darwin during his voyage on the HMS Beagle and the extensive botanical studies of Carl Linnaeus contributed to early ecological knowledge.

Medieval and Renaissance Naturalism

The Medieval and Renaissance periods witnessed the revival of interest in the natural world. Scholars like Leonardo da Vinci and Al-Jahiz made observations on ecological topics, including the roles of organisms in ecosystems. Leonardo da Vinci's ecological sketches and writings on the relationships between predators and prey demonstrate early ecological curiosity.

Indigenous Agricultural Practices

Indigenous agricultural practices, developed over millennia, reflect a deep understanding of ecological principles. Practices like crop rotation, agroforestry, and terracing are examples of sustainable land management.

These practices underscore the importance of traditional ecological knowledge in maintaining soil fertility, conserving water resources, and preserving biodiversity.

Transition to Modern Ecology

The transition from early ecological ideas to modern ecology occurred over centuries, marked by the contributions of scientists, naturalists, and philosophers. This period saw the emergence of ecological concepts like food chains, ecological niches, and the balance of nature.

Scientists like Charles Darwin, Alfred Russel Wallace, and Alexander von Humboldt made pivotal contributions to our understanding of species interactions, biogeography, and the interconnectedness of ecosystems.

Contemporary Relevance

Recognizing and appreciating the early roots of ecology is essential in understanding the holistic and interdisciplinary nature of contemporary ecological research. It emphasizes the continuity of ecological thought across cultures and time periods.

Early ecological insights continue to inform modern conservation efforts, sustainable practices, and our understanding of the intricate web of life on Earth.

The early roots of ecology are deeply intertwined with the intellectual curiosity and observations of ancient civilizations, philosophers, explorers, and indigenous communities. These early insights laid the foundation for the development of modern ecological thought and continue to inspire contemporary ecological research and conservation practices. Understanding this historical journey helps us appreciate the enduring and interconnected relationship between humans and the natural world.

3.3. RISE OF NATURAL HISTORY AND EXPLORATION

The "Rise of Natural History and Exploration" represents a crucial period in the history of ecology, where explorers, naturalists, and scientists embarked on voyages of discovery, cataloging and documenting the rich biodiversity of the world's ecosystems. This era laid the groundwork for modern ecological understanding and had a profound impact on the development of ecological thought. Here, we delve into this transformative period in ecological history:

Figure 12. Natural History and Exploration

(Source By ALBUQUERQUE, Luís de, pref – Livro das Armadas, pormenor [Academia das Ciências de Lisboa – ACIENL] (http://purl.pt/162/1/brasil/iconografia/armada/28_barco_cabral/index.html), Public Domain, https://commons.wikimedia.org/w/index.php?curid=2823827, Creative Commons License)

The Age of Exploration

The 15th to 17th centuries marked the Age of Exploration, characterized by European voyages of discovery to distant lands. These explorers encountered new environments, species, and ecosystems previously unknown to Western science.

Exploration expanded the known world and introduced European naturalists to the astonishing diversity of life on Earth. It fuelled curiosity about the natural world and laid the foundation for natural history.

Cataloguing Biodiversity

Early naturalists and explorers began systematically cataloging the flora and fauna they encountered during their travels. This period saw the emergence of botanical gardens, cabinets of curiosities, and the collection of specimens for study.

Naturalists like John Ray, Carolus Linnaeus, and Georg Wilhelm Steller made significant contributions to species classification and taxonomy during this era.

Taxonomy and Classification

The development of taxonomy, the science of naming and classifying organisms, became essential for organizing the growing body of knowledge about the diversity of life.

- Linnaean System: Carolus Linnaeus's binomial nomenclature system revolutionized taxonomy, providing a standardized way to name and categorize species, making it easier for scientists to communicate and compare findings.

Humboldtian Science

Alexander von Humboldt, an influential naturalist and explorer, pioneered a holistic approach to understanding nature. His work emphasized the interconnectedness of ecosystems, climate, and geography.

Humboldt's contributions laid the groundwork for biogeography, the study of the geographical distribution of species, and the recognition of the Earth as a dynamic, interconnected system.

Naturalists and Observations

Naturalists like Charles Darwin, Alfred Russel Wallace, and John Muir made groundbreaking observations during their travels. Darwin's voyage on the HMS Beagle and Wallace's work in Southeast Asia led to the development of the theory of evolution by natural selection.

Charles Darwin's insights into adaptation, species diversity, and the "struggle for existence" had profound implications for ecology, shaping our understanding of how species interact with their environments.

Encounters with Indigenous Knowledge

During their travels, explorers and naturalists often encountered Indigenous peoples with deep ecological knowledge. These encounters influenced their understanding of local ecosystems and the relationships between humans and nature.

Indigenous knowledge about plants, animals, and ecosystems was shared with Western naturalists, enriching ecological understanding and challenging Eurocentric views of nature.

Exploration's Impact on Conservation

The documentation of biodiversity and the recognition of the beauty and value of natural landscapes during this period had significant implications for the conservation movement.

The exploration era contributed to the establishment of national parks, wildlife reserves, and the protection of natural landscapes for future generations. Thus, the "Rise of Natural History and Exploration" represents a pivotal era in the history of ecology. It was marked by a spirit of discovery, the cataloging of biodiversity, the development of taxonomy, and groundbreaking ecological observations. This period laid the foundation for modern ecological thought, emphasizing the interconnectedness of life on Earth and the need for conservation and preservation. The legacy of exploration continues to influence ecological research and the broader understanding of our planet's natural wonders.

3.4. THE DARWINIAN REVOLUTION: CHARLES DARWIN'S IMPACT ON ECOLOGY AND EVOLUTIONARY THOUGHT

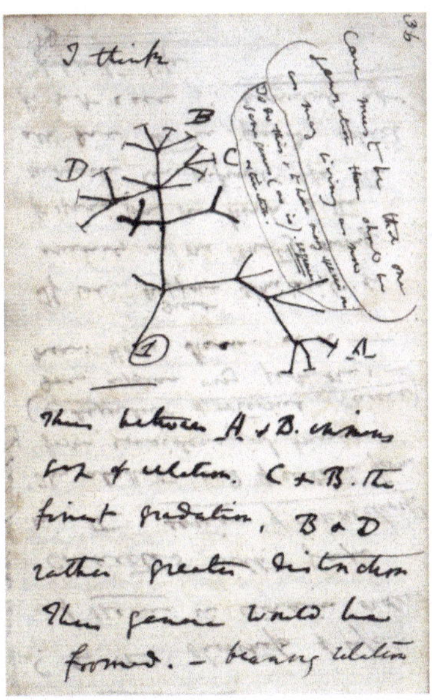

Figure 13. Charles Darwin's Thoughts on Ecology and Evolution

(Source: By Charles Darwin – Page 36 of Notebook B: [Transmutation of species (1837–1838)]. 'commenced. . . July 1837; from Darwin Online, Public Domain, https://commons.wikimedia.org/w/index.php?curid=36638808, Creative Commons License)

The Darwinian Revolution represents a transformative period in the history of ecology and evolutionary biology. It was driven by the groundbreaking work of Charles Darwin, particularly his publication of "On the Origin of Species" in 1859. This revolution reshaped our understanding of the natural world and provided profound insights into how species evolve and interact within ecosystems. Here, we delve into the Darwinian Revolution and its enduring significance:

Charles Darwin and "On the Origin of Species"

Charles Darwin's "On the Origin of Species" introduced the theory of evolution by natural selection. He proposed that species evolve over time through a process of differential survival and reproduction, driven by variations in traits.

Darwin's work emphasized the central role of adaptation, competition, and variation in shaping the diversity of life. His theory challenged traditional religious and creationist views and provided a naturalistic explanation for the complexity of life.

Evolutionary Ecology

Darwin's theory of evolution laid the foundation for evolutionary ecology. Ecologists began to study how natural selection and adaptation shape the behaviours, traits, and ecological interactions of species within ecosystems.

Research into topics like predator-prey relationships, mimicry, and coevolution became central to understanding the ecological dynamics of populations.

Concept of Ecological Niches

Darwin's ideas influenced the development of the concept of ecological niches. The notion that species occupy specific roles within ecosystems based on their adaptations and interactions with the environment became central to ecological thought.

Gause's Competitive Exclusion Principle: The work of ecologist Georgii Gause extended Darwin's ideas by proposing that species with identical niches cannot coexist indefinitely in the same ecosystem.

Biodiversity and Species Interactions

The Darwinian Revolution highlighted the importance of species interactions in ecological communities. Ecologists began to explore how competition, predation, mutualism, and parasitism influence species diversity and ecosystem structure.

Ecosystem Dynamics: The recognition of ecological relationships as dynamic processes shaped modern ecological theories and models, including those related to food webs and trophic interactions.

Contemporary Evolutionary Ecology

The legacy of the Darwinian Revolution endures in contemporary evolutionary ecology. Researchers continue to investigate topics like speciation, adaptation to changing environments, and the coevolutionary arms race between species.

Synthesis with Genetics: Modern evolutionary ecology incorporates genetics, genomics, and molecular biology to understand the genetic basis of adaptation and the mechanisms behind evolutionary change.

The Darwinian Legacy in Conservation

The concept of biodiversity conservation was influenced by Darwinian principles. Preserving genetic diversity within species, conserving ecosystems, and protecting keystone species all draw from Darwin's insights.

Conservation Biology: The field of conservation biology, which seeks to safeguard biodiversity and ecosystems, is built upon the understanding of evolution and ecological relationships, as elucidated by the Darwinian Revolution.

In conclusion, the Darwinian Revolution, propelled by Charles Darwin's groundbreaking work, reshaped our understanding of ecology and evolution. Darwin's theory of evolution by natural selection provided a unifying framework for explaining the diversity of life on Earth and the intricate interplay between species within ecosystems. His ideas continue to inspire and inform ecological research, conservation efforts, and our appreciation of the interconnectedness of all living organisms.

3.5. PLANT ECOLOGY AND SUCCESSION STUDIES: UNRAVELLING THE DYNAMICS OF ECOSYSTEMS

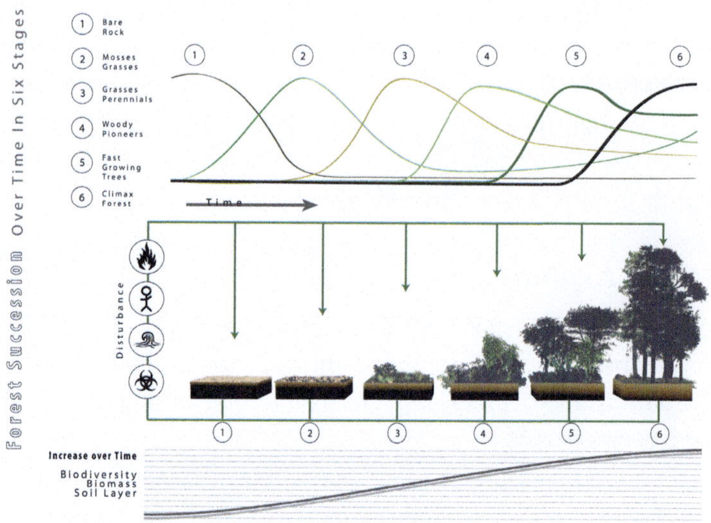

Figure 14. Plant Ecology and Succession

(Source: By LucasMartinFrey – Own work, CC BY 3.0, https://commons.wikimedia.org/w/index.php?curid=13274746 , Creative Commons License)

Plant ecology and succession studies represent a significant branch of ecological research that focuses on the interactions between plants, their environment, and the temporal changes in ecosystems. These studies provide crucial insights into the establishment, development, and maintenance of plant communities and their role in shaping ecosystems. Here, we explore the field of plant ecology and the study of ecological succession:

The Birth of Plant Ecology:

Plant ecology as a distinct field emerged in the late 19th and early 20th centuries. Pioneering ecologists like Frederic Clements and Henry Cowles conducted foundational research on the structure and dynamics of plant communities.

Clements introduced the concept of plant succession, which refers to the predictable sequence of plant species in a given area over time. Cowles' work on sand dune succession in the Indiana Dunes provided empirical evidence for ecological succession.

Ecological Succession:

Ecological succession is a fundamental concept in plant ecology that describes the gradual and predictable changes in a plant community following a disturbance or on a newly formed substrate.

Types of Succession: Primary succession occurs on bare, lifeless substrates (e.g., volcanic rock), while secondary succession occurs after a disturbance (e.g., forest fire or logging). The pioneer, intermediate, and climax stages are key components of succession.

Mechanisms: Succession is driven by a combination of factors, including changes in soil conditions, species interactions (e.g., competition and facilitation), and dispersal of seeds by wind, water, or animals.

Plant Adaptations and Traits:

Plant ecologists study the adaptations and traits that allow plants to thrive in specific environments. This includes adaptations for water conservation, nutrient acquisition, reproduction, and defense against herbivores.

Xerophytes, such as cacti, have evolved adaptations to conserve water in arid environments, while hydrophytes, like water lilies, have adaptations for life in aquatic habitats.

Community Ecology:

Plant ecologists investigate the interactions among plant species within communities. These interactions include competition for resources (e.g., light, water, and nutrients) and facilitation, where one species benefits another.

Diversity and Stability: Plant diversity within a community can enhance ecosystem stability and resilience. Studies examine the relationship between species richness and ecosystem functioning.

Human Impacts and Conservation:

Human activities, such as deforestation, urbanization, and agriculture, have profound effects on plant communities and ecosystems. Plant ecologists study the impacts of these activities and develop strategies for conservation and restoration.

Restoration Ecology: Restoration ecology focuses on restoring degraded ecosystems, including the reintroduction of native plant species and the removal of invasive species.

Global Change and Plant Responses:

Climate change and other global environmental changes are affecting plant distribution and behavior. Plant ecologists investigate how plants respond to changing environmental conditions and the implications for ecosystems and biodiversity.

Phenology: Studies on phenology, the timing of plant life cycle events like flowering and leaf budburst, help assess the impacts of climate change on plant behavior.

Contemporary Tools and Techniques

Advances in technology, including remote sensing, molecular biology, and ecological modeling, have revolutionized plant ecology. These tools enable researchers to study plant communities and their dynamics at various scales.

Big Data and Ecological Informatics: Plant ecologists use big data and informatics to analyze large datasets and model ecological processes.

Plant ecology and succession studies provide invaluable insights into the dynamics of plant communities and their interactions with the environment. They play a vital role in understanding how ecosystems respond to disturbances, adapt to changing conditions, and maintain biodiversity. These fields continue to evolve with new tools and approaches, offering critical guidance for conservation efforts and our understanding of the natural world.

3.6. THE BIRTH OF CONSERVATION ECOLOGY

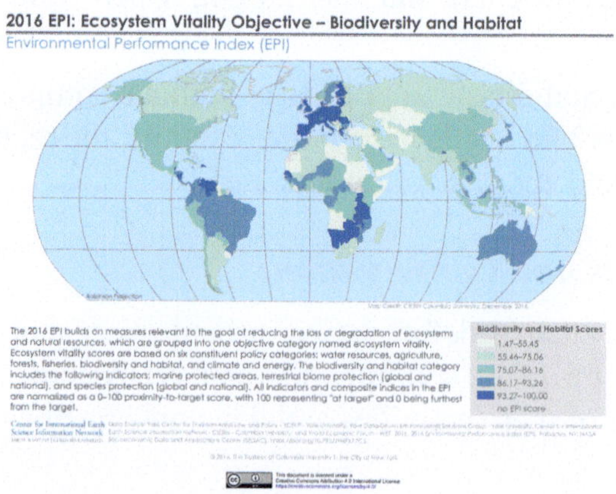

Figure 15. Birth of Conservation Ecology

(Source: By SEDACMaps – 2016 EPI: Ecosystem Vitality Objective – Biodiversity and Habitat, CC BY 2.0, https://commons.wikimedia.org/w/index.php?curid=85086688, Creative Commons License)

The birth of conservation ecology represents a pivotal moment in the history of environmental science and the understanding of humanity's impact on the natural world. This interdisciplinary field emerged in response to growing concerns about habitat loss, species extinction, and the degradation of ecosystems. Here, we explore the origins and significance of conservation ecology:

Historical Context

Conservation ecology emerged in the mid-20th century, during a period of increased awareness about the consequences of human activities on the environment.

The post-World War II era saw the rapid expansion of industrialization, urbanization, and agricultural intensification, leading to habitat destruction and pollution.

Early Conservation Movements

The birth of conservation ecology was closely linked to the conservation movements of the time, with pioneers advocating for the preservation of natural landscapes and the protection of endangered species.

Stalwarts like Aldo Leopold and Rachel Carson played influential roles in raising awareness about environmental issues. Leopold's book, "A Sand County Almanac," promoted the idea of a land ethic, while Carson's "Silent Spring" highlighted the dangers of pesticides.

Systematic Study of Biodiversity

Conservation ecology brought a systematic and scientific approach to the study of biodiversity. Researchers began documenting species distributions, population sizes, and habitat requirements.

Conservation ecologists emphasized the importance of biodiversity in maintaining ecosystem resilience, providing ecosystem services, and supporting human well-being.

Habitat Conservation and Restoration

One of the central themes of conservation ecology was the preservation and restoration of natural habitats. Researchers and conservationists worked to identify and protect critical ecosystems, such as wetlands, forests, and coral reefs.

Protected Areas: The establishment of national parks, wildlife reserves, and protected areas became a cornerstone of conservation efforts.

Conservation Genetics

Conservation ecology embraced genetic techniques to assess the health and genetic diversity of populations. This field, known as conservation genetics, aims to identify populations at risk of inbreeding and extinction.

Genetic Rescue: Conservation genetics also explored strategies such as genetic rescue, which involves introducing genes from healthy populations into endangered ones to increase genetic diversity.

Role of Legislation

Conservation ecology plays a crucial role in shaping environmental legislation and policies. The enactment of laws such as the Endangered Species Act (ESA) in the United States reflected the growing concern for protecting endangered species and their habitats.

International Agreements: Globally, conservation efforts led to international agreements like the Convention on International Trade in Endangered Species of Wild Fauna and Flora (CITES) and the Convention on Biological Diversity (CBD).

Conservation in the Anthropocene

The Anthropocene, characterized by widespread human influence on the planet, poses new challenges to conservation ecology. Researchers now grapple with issues like climate change, invasive species, and the ethical implications of conservation decisions.

Adaptive Management: Conservation ecologists have adopted adaptive management strategies to address dynamic and uncertain challenges in the Anthropocene.

Interdisciplinary Collaboration

Conservation ecology thrives on interdisciplinary collaboration, involving ecologists, geneticists, policymakers, economists, and social scientists.

It recognizes the interconnectedness of ecological, social, and economic systems.

Conservation Psychology: This subfield explores the psychological and behavioral aspects of conservation, aiming to understand and influence human attitudes and actions toward the environment.

The birth of conservation ecology marked a turning point in humanity's relationship with the natural world. It brought together scientific rigor, ethical considerations, and a commitment to safeguarding the Earth's biodiversity. Conservation ecology continues to evolve, adapt, and expand its scope to address the complex environmental challenges of the 21st century. It remains a critical discipline in our efforts to protect and sustain the planet's ecosystems and the species that inhabit them.

3.7 THE DEVELOPMENT OF ECOSYSTEM ECOLOGY: UNRAVELLING THE COMPLEX WEB OF LIFE

The development of ecosystem ecology represents a critical phase in the history of ecology. This subfield emerged in the mid-20th century and focused on understanding the complex interrelationships between organisms and their physical environments within ecosystems. Ecosystem ecology has provided valuable insights into nutrient cycling, energy flow, and the functioning of natural systems. Here, we explore the key milestones in the development of ecosystem ecology:

Early Ecological Foundations

Ecosystem ecology evolved from earlier ecological studies, including plant ecology, animal ecology, and studies of nutrient cycling. The works of early ecologists like Frederic Clements and Charles Elton contributed to the foundation of ecosystem thinking.

The Ecosystem Concept

The concept of an ecosystem was formally introduced in the mid-20th century. Ecosystems were defined as functional units consisting of living organisms (biotic components) and their physical and chemical environments (abiotic components). Ecosystem Pioneers: Scientists like Raymond Lindeman and Eugene Odum made pioneering contributions to ecosystem ecology. Lindeman's work on the flow of energy in ecosystems and Odum's development of ecological modeling were particularly influential.

Energy Flow and Trophic Levels

Ecosystem ecology introduced the concept of trophic levels, which represent the hierarchical positions of organisms in energy transfer within ecosystems. The flow of energy from producers to consumers and decomposers became a central focus.

Food Webs: Ecosystem ecologists studied food webs to depict the complex feeding relationships within ecosystems, showing how energy moves through different trophic levels.

Nutrient Cycling

Nutrient cycling, particularly the cycling of essential elements like carbon, nitrogen, and phosphorus, became a key aspect of ecosystem ecology. Research revealed how organisms take up, utilize, and release nutrients, influencing ecosystem productivity.

Biogeochemical Cycles: The study of biogeochemical cycles, such as the carbon cycle and nitrogen cycle, provided insights into how nutrients are recycled and redistributed within ecosystems.

Disturbance Ecology

Ecosystem ecologists recognized the role of disturbances, such as wildfires, hurricanes, and human activities, in shaping ecosystems. The field of disturbance ecology explores how ecosystems respond to and recover from disruptions.

Succession Studies: Succession, the predictable sequence of species in an ecosystem following a disturbance, was a critical focus within ecosystem ecology.

Human Impacts and Sustainability

Ecosystem ecology highlights the significant impacts of human activities on ecosystems, including habitat destruction, pollution, and overexploitation. This raised awareness about the need for sustainable resource management and conservation.

Ecosystem Services: The concept of ecosystem services emerged, emphasizing the benefits humans derive from healthy ecosystems, including clean water, pollination, and climate regulation.

Modern Tools and Techniques

Advances in technology, including remote sensing, isotopic analysis, and ecological modeling, revolutionized ecosystem ecology. These tools enable scientists to study ecosystems at various scales, from local to global.

Earth System Science: Ecosystem ecology has integrated with Earth system science, which examines the interactions between the geosphere, atmosphere, hydrosphere, and biosphere. This interdisciplinary approach broadens our understanding of Earth as a dynamic, interconnected system.

The development of ecosystem ecology has deepened our understanding of the intricate relationships between organisms and their environments. It has provided essential insights into the functioning of natural systems, nutrient cycling, energy flow, and the impacts of human activities on ecosystems. Ecosystem ecology remains a critical field in addressing global environmental challenges and promoting sustainable stewardship of the Earth's natural resources.

3.8. MODERN ENVIRONMENTALISM AND CONSERVATION BIOLOGY: FORGING A SUSTAINABLE FUTURE

Figure 16. Modern Environmentalism and Conservation Biology

(Source: By USPS, scanned by Greudin – USPS, Public Domain, https://commons.wikimedia.org/w/index.php?curid=70894, Creative Commons License)

Modern environmentalism and conservation biology are two closely intertwined fields that have emerged in response to growing concerns about environmental degradation, habitat loss, species extinction, and the need for sustainable resource management. These fields have played instrumental roles in shaping our understanding of ecological issues and driving efforts to protect and preserve the natural world. Here, we explore the key aspects and contributions of modern environmentalism and conservation biology:

Environmentalism as a Social Movement

Modern environmentalism can be traced back to the mid-20th century when it emerged as a social and political movement. Environmentalists advocated for increased awareness of environmental issues and the need for responsible stewardship of the planet.

Silent Spring: Rachel Carson's groundbreaking book "Silent Spring," published in 1962, is often credited with sparking the modern environmental movement. The book highlighted the harmful effects of pesticides, particularly DDT, on wildlife and ecosystems.

Founding Principles of Environmentalism

Environmentalism is founded on principles of environmental ethics, sustainability, and the recognition of the intrinsic value of nature. It emphasizes the interconnectedness of all life forms and the importance of safeguarding the environment for future generations.

Earth Day: The first Earth Day, held on April 22, 1970, marked a significant milestone in the environmental movement. It mobilized millions of people worldwide to raise awareness about environmental issues and advocate for environmental protection.

Emergence of Conservation Biology

Conservation biology is an interdisciplinary field that developed in response to the escalating threats to biodiversity and ecosystems. It incorporates principles from ecology, genetics, and environmental science to address conservation challenges.

Conservation biology was shaped by figures like Michael Soulé, George Schaller, and E.O. Wilson. Soulé's 1985 essay "What is Conservation Biology?" helped define the field's scope and objectives.

Biodiversity Conservation

Conservation biology places a strong emphasis on preserving biodiversity. It seeks to understand the factors driving species extinction, the importance of genetic diversity, and the creation of protected areas and wildlife corridors.

Keystone Species: The concept of keystone species, introduced by Robert T. Paine, highlights the critical role certain species play in maintaining ecosystem stability.

Ecosystem-Based Approaches

Conservation biology recognizes the importance of protecting entire ecosystems, including the services they provide to humans. Ecosystem-based approaches consider the interactions between species and the physical environment.

Restoration Ecology: This subfield focuses on restoring degraded ecosystems and habitats, often involving the reintroduction of native species and the removal of invasive species.

Applied Conservation Strategies

Conservation biology has contributed to the development of applied conservation strategies, such as captive breeding programs, habitat restoration, and species recovery plans.

Translocation and Reintroduction: Conservation biologists have successfully translocated and reintroduced endangered species into their native habitats to bolster populations.

Ethical Considerations and Environmental Policy

Both modern environmentalism and conservation biology grapple with ethical considerations, including questions of resource allocation, environmental justice, and the ethical treatment of animals.

These fields have influenced environmental policy and legislation, including the establishment of protected areas, wildlife conservation laws, and international agreements like the Convention on Biological Diversity (CBD).

Challenges in the Anthropocene:

Modern environmentalism and conservation biology face new challenges in the Anthropocene, an era characterized by human influence on the planet. These challenges include climate change, habitat fragmentation, and the need for sustainable resource management. To address these challenges, both fields increasingly collaborate with other scientific disciplines, policymakers, and local communities to develop holistic and effective conservation strategies.

Modern environmentalism and conservation biology represent our collective commitment to safeguarding the environment and preserving Earth's biodiversity. These fields have raised awareness about environmental issues, contributed to scientific understanding, and played pivotal roles in policy and conservation efforts. As we face complex challenges in the Anthropocene, the principles of environmentalism and the scientific rigor of conservation biology continue to guide our efforts toward a sustainable and harmonious relationship with the natural world.

3.9. CONTEMPORARY ECOLOGICAL CHALLENGES: NAVIGATING A CHANGING PLANET

Figure 17. Ecological Challenges

(Source: By Stephen Codrington, CC BY 2.5, https://commons.wikimedia.org/w/index.php?curid=221256, Creative Commons License)

Contemporary ecological challenges represent a complex and urgent set of issues that humanity faces in the 21st century. These challenges are characterized by the widespread influence of human activities on the environment, including climate change, habitat loss, biodiversity decline, and the degradation of ecosystems. Addressing these challenges requires interdisciplinary collaboration, innovative solutions, and a commitment to sustainable stewardship of the planet. Here, we explore some of the key contemporary ecological challenges:

Climate Change

Climate change, driven by the accumulation of greenhouse gases in the atmosphere, is one of the most pressing ecological challenges. Rising global temperatures lead to a cascade of impacts, including more frequent and severe heatwaves, extreme weather events, sea-level rise, and disruptions to ecosystems.

Mitigation and Adaptation: Strategies to address climate change include reducing greenhouse gas emissions through renewable energy adoption and carbon sequestration, as well as adapting to the changes that are already occurring.

Biodiversity Loss

Biodiversity is declining at an alarming rate due to habitat destruction, pollution, invasive species, overexploitation, and climate change. The loss of species and genetic diversity has profound ecological and societal consequences.

Conservation biology and protected areas play a critical role in preserving biodiversity. Conservation efforts also increasingly recognize the importance of local communities and indigenous knowledge.

Habitat Fragmentation

Habitat fragmentation occurs when natural landscapes are divided into smaller, isolated patches, often due to urbanization, agriculture, and infrastructure development. Fragmentation disrupts ecological processes, hinders species movement, and reduces genetic diversity.

Connectivity and Corridor Design: Conservation strategies include creating wildlife corridors and maintaining landscape connectivity to allow species to move and adapt to changing conditions.

Invasive Species

Invasive species, introduced by humans to new regions, can outcompete native species, disrupt ecosystems, and lead to biodiversity loss. Climate change can facilitate the spread of invasives to new areas.

Managing invasive species requires early detection, rapid response, and efforts to prevent further introductions. Biosecurity measures aim to reduce the risk of new invasions.

Pollution

Pollution from various sources, including industrial emissions, agriculture, and urban runoff, contaminates air, water, and soil. It poses health risks to ecosystems and human populations.

Environmental Regulations: Environmental regulations, such as emissions standards and waste disposal regulations, aim to mitigate pollution. Transitioning to sustainable practices can also reduce pollution.

Overexploitation of Resources

Overharvesting of natural resources, such as overfishing and deforestation, depletes ecosystems and threatens the long-term sustainability of these resources.

Sustainable Management: Sustainable resource management involves regulating harvest levels, enforcing fishing quotas, and promoting responsible forestry practices.

Land Use Change

Land use changes, including deforestation, urbanization, and agriculture expansion, alter landscapes and disrupt ecosystems. These changes contribute to habitat loss, carbon emissions, and altered water cycles.

Land Planning and Conservation: Sustainable land-use planning, reforestation, and the protection of critical habitats are essential for mitigating the negative impacts of land use change.

Emerging Diseases and Zoonoses

Emerging infectious diseases, including zoonotic diseases that transfer from animals to humans, pose global health and ecological threats. Factors like habitat destruction and wildlife trade increase the risk of zoonotic spillover events.

One Health Approach: A One Health approach involves collaboration between human, animal, and environmental health sectors to monitor and respond to disease threats at the interface of human and animal populations.

Ethics and Conservation Justice

Contemporary ecological challenges raise ethical questions about our responsibilities to other species and future generations. Conservation justice considers the equitable distribution of both the benefits and burdens of conservation efforts.

Inclusivity and Indigenous Knowledge: Solutions should involve diverse voices, including indigenous communities with valuable ecological knowledge, and address social and economic disparities.

Addressing contemporary ecological challenges requires a concerted global effort. It involves not only the application of science and technology but also a commitment to ethical principles, sustainable practices, and international cooperation. Meeting these challenges is essential for safeguarding the health of our planet and ensuring a sustainable and harmonious future for all life forms on Earth.

3.10. ROLE OF TECHNOLOGY AND BIG DATA IN ECOLOGY: REVOLUTIONIZING ENVIRONMENTAL UNDERSTANDING

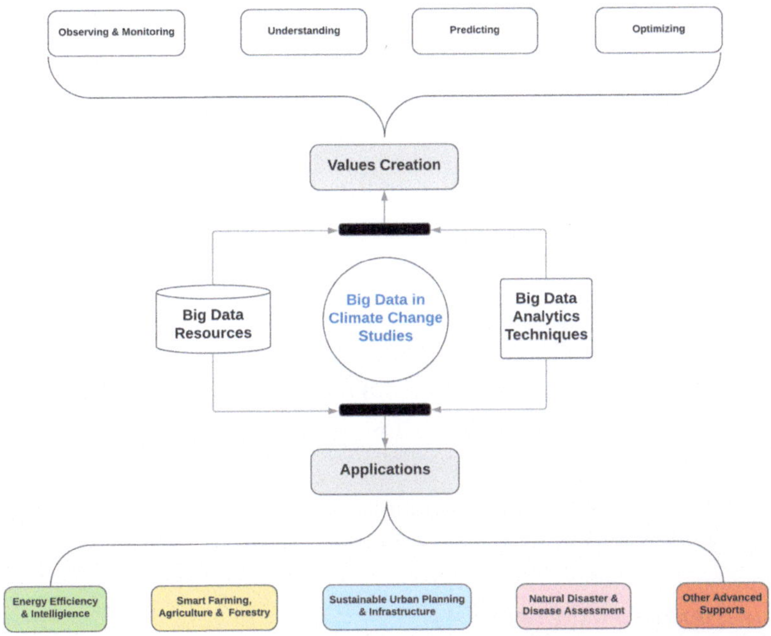

Figure 18. Role of Technology and Big Data in Ecology

(Source: Hassani, H., Huang X, Silva, E. Big Data and Climate Change. Big Data and Cognitive Computing. 2019; 3(1), 12. https://doi.org/10.3390/bdcc3010012, Creative Commons License)

Technology and big data have revolutionized the field of ecology, enabling researchers to collect, analyze, and interpret vast amounts of information about the natural world. These advances have enhanced our understanding of ecosystems, species interactions, and environmental change. Here, we delve into the pivotal role of technology and big data in ecology:

Remote Sensing and Earth Observation

Satellites equipped with remote sensing instruments provide valuable data on Earth's surface, including land cover, temperature, vegetation health, and ocean conditions. This information is critical for monitoring large-scale ecological processes and environmental changes.

Remote sensing technology helps track deforestation, monitor wildfires, assess crop health, and study changes in polar ice cover. It also aids in wildlife conservation and habitat mapping.

GPS and Tracking Technologies:

Global Positioning System (GPS) and radio telemetry enable researchers to track the movements of animals with unprecedented accuracy. These technologies provide insights into migration patterns, habitat use, and behavior.

GPS tracking has been used to monitor the movements of large mammals like elephants and whales, aiding in conservation efforts and understanding their ecological roles.

Camera Traps and Remote Imaging

Camera traps and remote imaging devices capture images and videos of wildlife in their natural habitats. These non-invasive tools help researchers monitor elusive and nocturnal species.

Biodiversity Monitoring: Camera traps have contributed to biodiversity assessments, population estimates, and the discovery of new species.

Environmental DNA (eDNA)

eDNA analysis involves extracting and analyzing DNA shed by organisms into their environment, such as water or soil. This technology allows for the detection of species and their presence without direct observation. Aquatic Ecosystems: eDNA has been particularly useful in aquatic ecosystems, where it can detect the presence of rare or invasive species.

Acoustic Monitoring

Acoustic sensors and underwater microphones, known as hydrophones, record sounds in natural environments. These recordings help researchers monitor animal vocalizations and behavior.

Marine Mammals: Acoustic monitoring is essential for studying marine mammal populations, including whale songs and dolphin communication.

DNA Sequencing and Genomics

Advances in DNA sequencing and genomics have revolutionized the study of biodiversity. Researchers can analyze the genetic makeup of species, track population dynamics, and uncover evolutionary relationships.

Conservation Genetics: DNA analysis informs conservation efforts, such as identifying unique genetic lineages and assessing the genetic health of endangered populations.

Big Data Analytics

Big data analytics involves processing and analyzing large datasets to extract meaningful patterns and insights. In ecology, this approach is used to understand complex ecological interactions and trends.

Ecological Models: Big data feeds ecological models that predict species distributions, habitat suitability, and responses to environmental changes.

Citizen Science and Data Crowdsourcing

Citizen science initiatives involve the public in ecological data collection. Technology facilitates data crowdsourcing, enabling people to contribute observations and data.

Projects like eBird for birdwatching and iNaturalist for biodiversity observations harness the collective power of citizen scientists.

Challenges and Ethical Considerations

While technology and big data offer transformative opportunities, they also raise ethical concerns related to data privacy, the impact of sensor networks on ecosystems, and the equitable distribution of benefits and access.

Ethical Data Use: The responsible use of technology and big data in ecology includes respecting privacy, ensuring data security, and promoting transparency in data sharing.

In summary, technology and big data have become integral to modern ecological research and conservation efforts. These tools provide unprecedented access to environmental information, allowing scientists to address complex ecological challenges and make informed decisions for the protection and management of ecosystems and species. Balancing the benefits of technology with ethical considerations is crucial as we continue to advance our understanding of the natural world.

3.11. INTERDISCIPLINARY APPROACHES IN MODERN ECOLOGY: BRIDGING KNOWLEDGE FOR HOLISTIC UNDERSTANDING

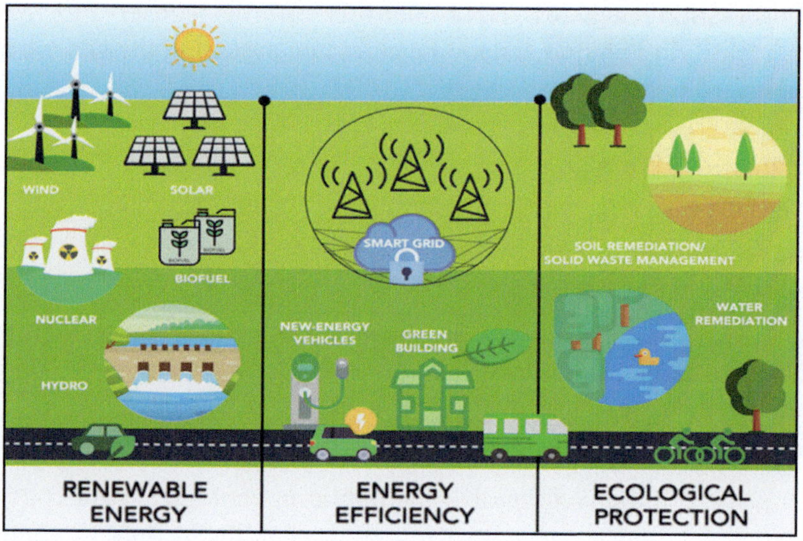

Figure 19. Interdisciplinary Approaches in Modern Ecology

(Source: Xu, X., Sun, Y., Krishnamoorthy, S., & Chandran, K. An Empirical Analysis of Green Technology Innovation and Ecological Efficiency Based on a Greenhouse Evolutionary Ventilation Algorithm Fuzzy-Model. Sustainability. 2020; 12(9): 3886. https://doi.org/10.3390/su12093886, Creative Commons License)

Modern ecology has increasingly embraced interdisciplinary approaches, recognizing that complex ecological challenges require diverse perspectives and expertise to be effectively addressed. These approaches involve collaboration between ecologists and experts from various fields to gain a more holistic understanding of ecosystems, biodiversity, and the environment. Here, we explore the role and significance of interdisciplinary approaches in modern ecology:

Ecological Economics

Ecological economics combines principles of ecology and economics to assess the ecological sustainability of economic systems. It considers the environmental costs of economic activities and explores alternative models of sustainable development.

Ecological economics helps evaluate the value of ecosystem services, assess the impact of human activities on ecosystems and design policies for sustainable resource management.

Environmental Sociology

Environmental sociology examines the social and cultural factors influencing human interactions with the environment. It explores how societal values, norms, and institutions shape environmental behaviors and policies.

Environmental sociology informs conservation efforts, environmental education, and the design of communication strategies to promote sustainable behaviors.

Conservation Psychology

Conservation psychology investigates the psychological and emotional connections between individuals and the natural world. It explores the factors that influence pro-environmental behaviors and attitudes.

Insights from conservation psychology are used in environmental education, ecotherapy, and strategies to promote environmental stewardship.

Landscape Ecology

Landscape ecology considers the spatial arrangement of habitats and ecosystems. It examines how landscape patterns influence ecological processes, species distributions, and biodiversity.

Landscape ecology informs land-use planning, habitat conservation, and strategies for mitigating the effects of habitat fragmentation.

Evolutionary Ecology

Evolutionary ecology integrates principles of evolutionary biology and ecology to understand how species adapt to their environments and how ecological interactions drive evolutionary processes.

This approach helps explain the coevolution of species, the development of adaptive traits, and the role of ecological interactions in speciation.

Conservation Genetics

Conservation genetics combines genetics and ecology to study the genetic diversity and health of populations. It assesses the impact of genetic factors on species viability and adaptability.

Conservation genetics informs breeding programs, population management, and the identification of genetically distinct populations for conservation.

Geospatial Analysis and GIS

Geographic Information Systems (GIS) and geospatial analysis use spatial data to model and analyze ecological patterns, habitat suitability, and species distributions.

GIS aids in conservation planning, habitat mapping, and the identification of biodiversity hotspots.

Earth System Science

Earth system science takes an interdisciplinary approach to studying Earth as a complex, integrated system. It explores the interactions between the atmosphere, hydrosphere, biosphere, and geosphere.

This approach contributes to our understanding of climate change, biogeochemical cycles, and the impacts of human activities on Earth's systems.

Interdisciplinary Research Centers

Many research institutions and universities establish interdisciplinary centers and programs dedicated to ecological research, fostering collaboration between ecologists, engineers, social scientists, and other experts.

Collaboration Hubs: These centers serve as hubs for interdisciplinary research, offering resources, expertise, and funding opportunities to address complex ecological challenges. Therefore, interdisciplinary approaches in modern ecology break down traditional boundaries and silos, fostering collaboration and a deeper understanding of the intricate relationships between ecosystems and the broader environment. They empower researchers and decision-makers to develop more comprehensive and effective strategies for conserving biodiversity, mitigating environmental degradation, and promoting sustainable stewardship of our planet.

3.12. FUTURE DIRECTIONS IN ECOLOGY: NAVIGATING EMERGING CHALLENGES AND OPPORTUNITIES

Ecology, as a field, is continually evolving to address new challenges and opportunities presented by a changing world. As we look to the future, several key directions and priorities are likely to shape the trajectory of ecological research and its applications. Here, we explore some of the future directions in ecology:

Climate Change Ecology

Climate change is a defining challenge of our era, and ecology will play a crucial role in understanding and mitigating its impacts. Future research will focus on how ecosystems respond to changing climate patterns, including shifts in species distributions, altered phenology, and adaptations to temperature and precipitation changes.

Ecology will inform climate adaptation strategies, such as assisted migration of species, reforestation efforts, and the development of resilient agricultural practices.

Urban Ecology

As the world becomes increasingly urbanized, urban ecology will grow in importance. Future research will explore the ecological dynamics of cities, including the effects of urbanization on biodiversity, urban heat islands, and ecosystem services in urban environments.

Urban ecology will guide sustainable urban planning, green infrastructure development, and strategies to enhance the quality of life in cities while minimizing ecological impacts.

Conservation in the Anthropocene

The Anthropocene, characterized by human influence on the planet, presents unique conservation challenges. Future research will address issues such as invasive species management, novel ecosystems, and the ethical dimensions of conservation in a human-dominated world.

Conservation efforts will need to adapt to the realities of the Anthropocene by integrating human needs and ecological goals, promoting coexistence with wildlife, and exploring rewilding initiatives.

Ecosystem Restoration and Resilience

Restoration ecology will continue to gain prominence as ecosystems face increasing threats. Future research will focus on techniques for restoring degraded habitats, enhancing ecosystem resilience, and evaluating the long-term success of restoration efforts.

Ecosystem restoration will be critical for recovering biodiversity, improving ecosystem services, and mitigating the impacts of habitat loss and fragmentation.

Biodiversity Monitoring and Big Data

Advances in technology and big data analytics will transform biodiversity monitoring. Future research will involve the use of remote sensing, environmental DNA, and citizen science to track changes in biodiversity and ecosystems at unprecedented scales.

Real-time biodiversity monitoring will inform conservation decision-making, help detect early warning signs of ecological disruptions, and support evidence-based policy development.

Global Collaborations and Conservation Networks

The interconnectedness of ecological challenges demands global collaboration. Future research will involve international partnerships, data sharing, and coordinated efforts to address issues such as migratory species conservation, transboundary pollution, and ecosystem restoration.

Collaborations and networks will facilitate the exchange of knowledge, resources, and best practices, leading to more effective conservation strategies on a global scale.

Interdisciplinary Approaches

Ecology will increasingly integrate with other fields, such as economics, sociology, and public health, to address complex environmental challenges. Interdisciplinary approaches will provide holistic solutions to issues like biodiversity loss, ecosystem services valuation, and human-wildlife conflicts.

These approaches will inform policy development, resource management, and sustainable practices by considering ecological, social, and economic dimensions.

Environmental Ethics and Education

As ecological challenges become more pronounced, the ethical dimensions of environmental stewardship will receive greater attention. Future research will explore questions of ethics, environmental justice, and the responsibilities of individuals and societies in preserving the environment.

Environmental education will play a crucial role in shaping the values and behaviours needed for sustainable living and responsible environmental decision-making.

Hence, the future of ecology is marked by a commitment to addressing complex, interconnected challenges while seizing opportunities to better understand and protect the natural world. Collaboration, innovation, and a deep appreciation for the ecological principles that underpin our planet's health will guide the discipline as it continues to evolve and contribute to a sustainable and harmonious future.

3.13. HISTORY OF HUMAN ECOLOGY: UNDERSTANDING HUMANITY'S RELATIONSHIP WITH THE ENVIRONMENT

Human ecology, as a field of study, emerged to explore the intricate relationship between human societies and their environments. This interdisciplinary field draws from ecology, anthropology, sociology, geography, and other disciplines to examine how humans interact with and impact the natural world. Here, we explore the history and evolution of human ecology:

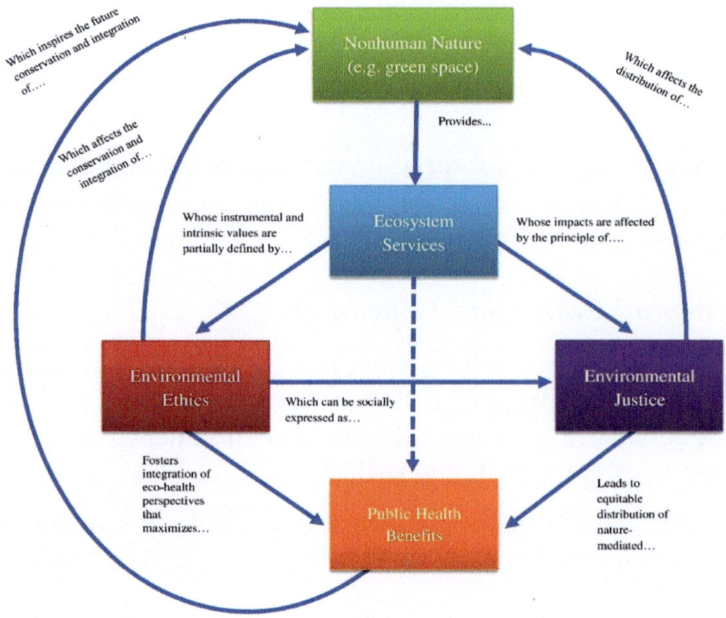

Figure 20. Human Ecology

(Source: By Viniece Jennings, Jessica Yun, Lincoln Larson – Viniece Jennings et. al. "Finding Common Ground: Environmental Ethics, Social Justice, and a Sustainable Path for Nature-Based Health Promotion," Healthcare doi:10.3390/healthcare4030061, CC BY 4.0, https://commons.wikimedia.org/w/index.php?curid=75480372, Creative Commons License)

Early Roots

Human ecology has ancient roots, with early societies adapting to their environments and ecosystems. Indigenous knowledge systems and practices often reflect sustainable interactions with nature.

Indigenous communities worldwide developed sustainable agricultural methods, such as crop rotation, and maintained harmonious relationships with local ecosystems.

Early Anthropological Contributions

Anthropologists in the late 19th and early 20th centuries, such as Franz Boas and Bronisław Malinowski, began examining how culture and environment influenced human societies.

- Environmental Determinism: Some early anthropologists advocated for environmental determinism, the idea that climate and geography strongly shape human cultures and behaviors.

Chicago School of Sociology

The Chicago School of Sociology, in the early 20th century, played a pivotal role in shaping the field of human ecology. Scholars like Robert Park and Ernest Burgess emphasized the social aspects of urban environments.

- Urban Ecology: This approach explored how cities and urbanization influenced human behaviors, social interactions, and living conditions.

The Systems Approach

Human ecology began adopting a systems approach in the mid-20th century. This approach examined the complex interactions between humans and their environments, emphasizing feedback loops and interconnectedness.

Human ecologists studied how changes in one component of a system (e.g., population growth or technology) could lead to broader environmental and social consequences.

Ecological Anthropology

Ecological anthropology emerged as a subfield that focused on the relationships between cultures and their environments. It incorporated elements of cultural anthropology and ecology to understand how societies adapt to ecological constraints.

- Cultural Ecology: Cultural ecologists, such as Julian Steward, explored how cultural practices and institutions adapt to ecological conditions.

Political Ecology

Political ecology, which gained prominence in the late 20th century, examined the political and economic dimensions of environmental issues. It explored how power dynamics, resource distribution, and environmental degradation intersect.

- Environmental Justice: Political ecology contributed to the understanding of environmental justice issues, highlighting how marginalized communities often bear the brunt of environmental impacts.

Contemporary Human Ecology

Modern human ecology addresses pressing global challenges, including climate change, biodiversity loss, and sustainability. It incorporates advances in technology and the integration of ecological principles with social sciences.

- Sustainability Science: Human ecology now plays a critical role in sustainability science, aiming to find solutions to complex environmental and societal problems.

Future Directions

The future of human ecology lies in addressing pressing issues, such as sustainable resource management, climate adaptation, and the ethical dimensions of human-environment interactions.

- Interdisciplinary Collaboration: Collaboration with other fields, including environmental economics, public health, and policy studies, will be essential to finding comprehensive solutions to global challenges.

Today, human ecology represents a dynamic and evolving field that seeks to understand and address the complex interplay between human societies and the environment. It draws from a rich history of interdisciplinary scholarship and continues to adapt and expand its scope to tackle the environmental challenges of the 21st century. As humanity grapples with issues of sustainability and environmental stewardship, human ecology will remain a crucial discipline in shaping our relationship with the natural world.

3.14. REFERENCES

1. Bocking, Stephen. (1997). Ecologists and Environmental Politics: A History of Contemporary Ecology. Yale University Press, New Haven.
2. Croker, Robert A. (1991). Pioneer Ecologist: The Life and Work of Victor Ernest Shelford, 1877–1968. Washington, D.C.
3. Egerton, Frank N. (ed.). (1977). History of American Ecology. Arno Press, New York.
4. Egerton, Frank N. (2012). Roots of Ecology: Antiquity to Haeckel. University of California Press, Berkeley.
5. Golley, Frank B. (1993). A History of the Ecosystem Concept: More than the Sum of the Parts. Yale University Press, New Haven.
6. Kingsland, Sharon E. (2005). The Evolution of American Ecology 1890–2000. The Johns Hopkins University Press, Baltimore.
7. Levin, Simon A. (2010). The evolution of ecology. Chronicle of Higher Education 56(42), B9–B11.
8. McIntosh, Robert P. (1985). The Background in Ecology: Concept and Theory. Cambridge University Press, Cambridge.
9. Odum, Eugene P. (1977). The emergence of ecology as a new integrative discipline. Science 15, 1289–1293.
10. Park, Orlando. (1945). Observations concerning the future of ecology. Ecology 26, 1–9.
11. Pearsall, W. H. (1964). The development of ecology in Britain. Journal of Ecology 52 (Supplement):1–12.
12. Real, Leslie A., & James H. Brown, (eds.). Foundations of Ecology. University of Chicago Press, Chicago.
13. Salisbury, Sir Edward. (1964). The origin and early years of the British Ecological Society. Journal of Ecology 52 (Supplement):13–18.
14. Sears, Paul B. (1969). Plant Ecology. pp. 124–131. In J. Ewan, ed. A Short History of Botany in the United States. Hafner, New York.
15. Worster, Donald. (1994). Nature's Economy: A History of Ecological Ideas. Cambridge University Press, Cambridge.

CHAPTER 4

General Ecology Concepts: Understanding the Foundations of Ecosystems and Biodiversity

CONTENTS

4.1. INTRODUCTION

Ecology is the scientific study of how living organisms interact with each other and their environment. It encompasses a wide range of concepts that help ecologists understand the complex relationships between organisms, their surroundings, and the broader ecosystems they inhabit. Here are some fundamental ecology concepts:

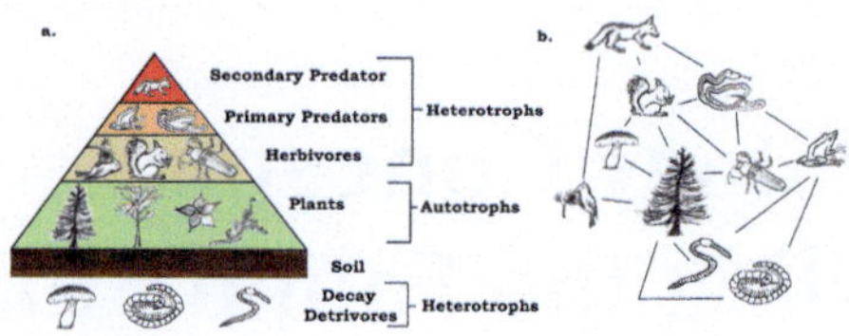

Figure 21. Ecology Concepts

(Source: By Thompsma – Own work, CC BY 3.0, https://commons.wikimedia.org/w/index.php?curid=15367675, Creative Commons License)

Ecosystem

An ecosystem is a self-contained, functional unit consisting of living organisms (biotic components) and their physical and chemical environment (abiotic components). Ecosystems can vary in size, from small ponds to entire forests.

Ecosystems exhibit energy flows and nutrient cycling among their components, and they are characterized by specific biotic communities adapted to their environmental conditions.

Biotic and Abiotic Factors

Biotic factors are the living components of an ecosystem, including plants, animals, fungi, and microorganisms. Abiotic factors are non-living components, such as temperature, water availability, soil composition, and sunlight.

Biotic and abiotic factors interact to shape the structure and function of ecosystems. Organisms adapt to and influence these factors in various ways.

Species

A species is a group of organisms that can interbreed and produce fertile offspring. Each species has a unique set of traits and adaptations.

Species are the building blocks of biodiversity, and their interactions drive ecological processes like predation, competition, and mutualism.

Population

A population is a group of individuals of the same species living in a particular area at a specific time.

Population ecology studies the factors that influence population size, growth, distribution, and density over time.

Community

A community consists of all the populations of different species living and interacting in a particular area.

Community ecology explores the interactions between species, including competition, predation, mutualism, and parasitism.

Habitat and Niche

Habitat: A habitat is the specific physical and biological environment in which an organism lives.

Niche: A niche refers to the role an organism plays within its habitat, including its interactions with other species and how it acquires and uses resources.

Food Web and Trophic Levels

Food Web: A food web is a complex network of feeding relationships among species in a community. It illustrates who eats whom in an ecosystem.

Trophic Levels: Trophic levels categorize organisms based on their position in the food chain. Producers (plants) are at the base, followed by herbivores, primary carnivores, and so on.

Energy Flow and Nutrient Cycling

Energy Flow: Energy flows through ecosystems, moving from producers to consumers and decomposers. Only a portion of energy is transferred from one trophic level to the next.

Nutrient Cycling: Nutrients like carbon, nitrogen, and phosphorus cycle through ecosystems, being taken up by organisms, released back into the environment during decomposition, and reused by other organisms.

Succession

Succession is the process by which communities of organisms change over time, typically following a disturbance. It can be primary (starting from bare rock or soil) or secondary (after a disturbance like a fire).

Climax Community: Succession often leads to the development of a climax community, which is a stable and mature ecosystem that remains relatively unchanged unless disturbed.

Biodiversity

Biodiversity refers to the variety of life forms, including species diversity, genetic diversity within species, and ecosystem diversity.

Biodiversity contributes to ecosystem stability, resilience, and the provision of ecosystem services that benefit humans.

These are foundational concepts in ecology that provide a framework for understanding the dynamics of life on Earth and the intricate relationships between organisms and their environments. Ecologists use these concepts to investigate ecological patterns and processes, assess environmental changes, and inform conservation and management strategies.

4.2. ECOLOGICAL SUCCESSION: UNDERSTANDING THE DYNAMIC CHANGE OF ECOSYSTEMS

Ecological succession is a fundamental concept in ecology that describes the process of how ecosystems change and develop over time in response to environmental disturbances, such as fires, floods, human activities, or the colonization of new terrain. This dynamic process involves a sequence of predictable and often orderly changes in the composition and structure of biological communities within an ecosystem. Here, we delve into the concept of ecological succession:

Primary Succession

Primary succession occurs on bare, lifeless substrates devoid of soil, such as newly formed volcanic islands, bare rock surfaces, or glacial moraines.

Primary succession begins with the colonization of pioneer species, often lichens and mosses, which can tolerate harsh conditions. These early colonizers break down rock and organic matter, gradually creating soil. Over time, more complex plant species, such as grasses and shrubs, establish themselves, leading to the development of a stable, climax community. The gradual formation of soil and the establishment of plants on lava flows provide an example of primary succession.

Secondary Succession

Secondary succession occurs in areas where an existing ecosystem has been disturbed or partially destroyed but still retains some soil and seeds.

Secondary succession begins with the rapid colonization of pioneer species that can quickly exploit available resources. As these early colonizers grow, they create favorable conditions for other plant species to establish themselves. Over time, a more complex and diverse community develops, ultimately reaching a climax state.

Forests regrowing after a wildfire or abandoned agricultural fields reverting to natural vegetation are examples of secondary succession.

Pioneer Species

Pioneer species are the first organisms to colonize a newly disturbed or barren area.

Pioneer species play a crucial role in primary and secondary succession by breaking down rock, enriching the soil with organic matter, and creating conditions that are more favorable for other species to establish themselves.

Climax Community

A climax community is the stable and relatively unchanging end point of succession for a particular ecosystem. It represents a state of ecological equilibrium where the community is in balance with its environment.

Climax communities are typically characterized by a diverse array of species well-adapted to the prevailing environmental conditions. They are more resilient to disturbances and exhibit complex species interactions.

Disturbance

Disturbance refers to any event or factor that disrupts an ecosystem, potentially leading to ecological succession.

Disturbances can be natural (e.g., wildfires, hurricanes, volcanic eruptions) or human-induced (e.g., deforestation, urbanization, agriculture).

Facilitation, Inhibition, and Tolerance

Facilitation: Some species, particularly pioneer species, create conditions that facilitate the establishment of other species. For example, nitrogen-fixing plants can enrich the soil with nutrients, benefiting neighboring plants.

Inhibition: Some early colonizers may inhibit the establishment of other species through competition or allelopathy (chemical inhibition).

Tolerance: Some species are tolerant to harsh environmental conditions and can establish themselves even in the presence of other species.

Ecological Resilience

Ecological resilience refers to the capacity of an ecosystem to withstand disturbances, adapt to changing conditions, and recover to its previous state or transition to a new state.

Understanding succession and resilience is essential for ecosystem management, conservation, and restoration efforts.

Ecological succession is a dynamic and ongoing process that shapes the structure and composition of ecosystems. It highlights the resilience and adaptability of nature in responding to environmental changes and disturbances, ultimately leading to the formation of diverse and stable biological communities.

4.2.1. Primary Succession

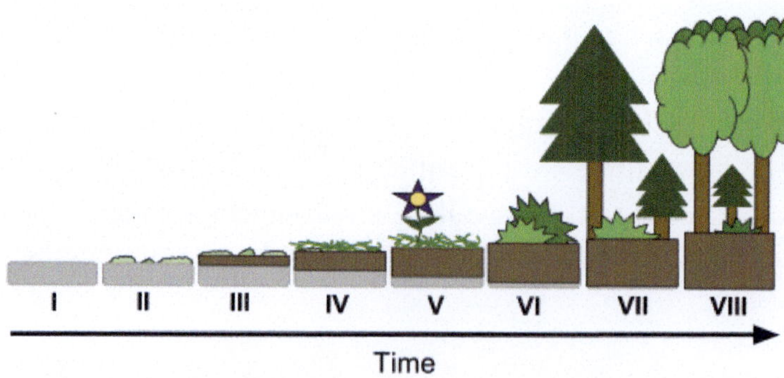

Figure 22. Primary Succession

(Source: https://en.m.wikipedia.org/wiki/File:AP_Biology_-_Primary_Succession_Drawing.svg, Creative Commons License)

Pioneer Species

Pioneer species are the first organisms to colonize a previously lifeless or barren environment in the early stages of primary succession.

Pioneer species are typically hardy and well-adapted to harsh conditions, including poor soil, extreme temperatures, and limited water availability. They often include lichens, mosses, and certain plant species that can establish themselves on rocky or barren substrates.

Pioneer species play a critical role in primary succession by initiating the process of soil formation. They can break down rocks, stabilize the substrate, and accumulate organic matter as they grow and die. This organic matter gradually enriches the soil, making it more suitable for other, less hardy plant species to colonize.

Ruderal Species

Ruderal species are a group of plants that are adapted to disturbed and often man-made environments. They are characterized by their ability to thrive in areas with high levels of disturbance, such as abandoned urban lots, roadsides, or recently cleared land.

Ruderal species typically have traits that allow them to rapidly colonize and reproduce in open, disturbed habitats. These traits may include quick germination, high seed production, and the ability to tolerate a wide range of environmental conditions.

Ruderal species are often among the first to appear in secondary succession, particularly in ecosystems recovering from human activities or recent disturbances. They can establish themselves in areas where the soil has been exposed or disturbed and often act as pioneer species in such environments.

Supertramp

The term "supertramp" is not a commonly used ecological term. It may refer to a concept or idea in a specific ecological context, but it does not have a standard definition in mainstream ecology.

In some cases, "supertramp" could be used informally to describe a species that exhibits extraordinary dispersal capabilities, allowing it to colonize new habitats rapidly. Such species might have unique adaptations or behaviors that make them highly effective at long-distance dispersal.

So, pioneer species are the initial colonizers of barren environments in primary succession, ruderal species are adapted to thrive in disturbed habitats and often act as pioneers in secondary succession, and "supertramp" may refer to a concept related to highly dispersal-capable species but lacks a standard ecological definition.

4.2.2. Secondary Succession: The Rebuilding of Ecosystems after Disturbance

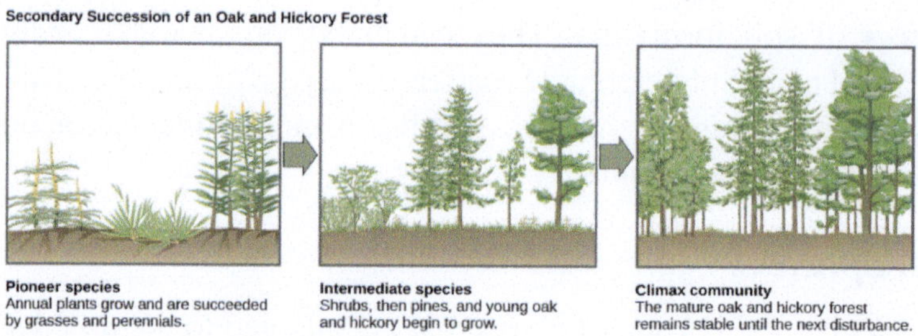

Figure 23. Secondary Succession

(Source: By CNX OpenStax – http://cnx.org/contents/GFy_h8cu@10.53:rZudN6XP@2/Introduction, CC BY 4.0, https://commons.wikimedia.org/w/index.php?curid=49938778, Creative Commons License)

Secondary succession is a critical ecological process that occurs when an existing ecosystem has been disturbed, damaged, or altered, but the soil and some biotic components remain intact. Unlike primary succession, which starts on entirely barren substrates, secondary succession begins in areas that have a history of supporting life. Here's a detailed overview of secondary succession:

Definition

Secondary succession is the process of ecological development that occurs after a disturbance has disrupted or partially destroyed an existing ecosystem. It involves the gradual return of life to an area where some soil and seeds or propagules of plants are still present.

Disturbances

Secondary succession can be triggered by various disturbances, including natural events like wildfires, storms, or volcanic eruptions, as well as human-induced activities like deforestation, farming abandonment, or construction.

The severity of the disturbance can vary widely, ranging from a localized disturbance like a fallen tree to more extensive events like a forest fire.

Stages of Secondary Succession

Pioneer Species: Secondary succession often begins with pioneer species, which are typically fast-growing and well-adapted to colonize open, disturbed habitats. These species can include grasses, shrubs, or herbaceous plants.

Soil Development: As pioneer species grow and die, they contribute organic matter to the soil, helping to improve its fertility and structure. This facilitates the establishment of a wider range of plant species.

Intermediate Succession: Over time, more competitive plant species may replace the pioneers, leading to a process called intermediate succession. These species may be taller, shade-tolerant, or have more specialized adaptations.

Climax Community: Eventually, the process of succession leads to the establishment of a climax community. This community represents a relatively stable and diverse array of species that is well-suited to the prevailing environmental conditions. The climax community is capable of persisting for an extended period without significant changes unless a new disturbance occurs.

Timeline

Secondary succession occurs over a variable timeline, depending on factors such as the severity of the disturbance, local climate, and the availability of propagules (seeds or spores). It can take decades or even centuries for an ecosystem to reach a climax state.

Ecological Importance

Habitat Restoration: Secondary succession is a vital process in habitat restoration efforts. Natural or human-induced disturbances can create opportunities for re-establishing native vegetation and promoting biodiversity. Ecosystem Services: Understanding secondary succession is crucial for assessing the capacity of ecosystems to recover and provide essential ecosystem services, such as carbon sequestration, water purification, and habitat provision.

Human Impact

Human activities, including land-use changes, deforestation, and urbanization, can disrupt ecosystems and trigger secondary succession. Understanding how these activities affect ecological recovery is essential for sustainable land management and conservation.

In summary, secondary succession is a dynamic process by which ecosystems recover and rebuild after disturbance. It involves a series of stages, from pioneer species colonizing disturbed areas to the development of a diverse climax community. Understanding secondary succession is crucial for ecological restoration, land management, and conservation efforts in a world where ecosystems are continually influenced by natural and human-induced disturbances.

4.3. CARRYING CAPACITY AND ECOLOGICAL COLLAPSE

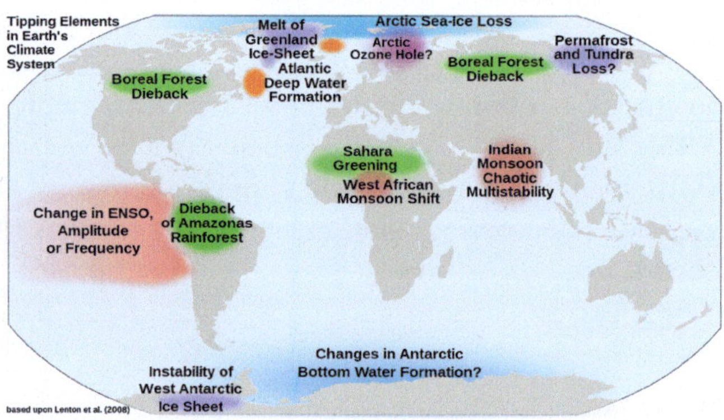

Figure 24. Carrying Capacity and Ecological Collapse

(Source: By CodeOne (blank map), DeWikiMan (additional elements) – This file was derived from: WorldMap.svgThe content is based upon Figure 2 in Kriegler et al. (2009) and Lenton et al. (2008)., CC BY-SA 4.0, https://commons.wikimedia.org/w/index.php?curid=74405407, Creative Commons License)

Carrying capacity and ecological collapse are interconnected concepts in ecology that relate to the sustainable utilization of resources and the consequences of exceeding the capacity of an ecosystem to support its inhabitants. Let's explore these concepts in detail:

Carrying Capacity

Carrying capacity refers to the maximum number of individuals of a particular species that an ecosystem can support over the long term without causing environmental degradation or resource depletion.

Several factors determine the carrying capacity of an ecosystem, including the availability of resources (e.g., food, water, habitat), environmental conditions (e.g., temperature, precipitation), and interactions with other species (e.g., competition, predation).

Carrying capacity is not a fixed value but can vary over time due to changes in environmental conditions, resource availability, and the presence of other species.

Understanding carrying capacity is essential for sustainable resource management, conservation efforts, and the prevention of overexploitation of natural resources. It helps set sustainable harvest limits for fisheries, wildlife populations, and agricultural systems.

Ecological Collapse

Ecological collapse occurs when an ecosystem experiences a rapid and drastic decline in its structure and function, leading to a loss of biodiversity, disruption of ecological processes, and a loss of ecosystem services.

Ecological collapse can result from a variety of factors, including overexploitation of resources, habitat destruction, pollution, invasive species, climate change, and alterations to key ecological processes.

Ecological collapse can have far-reaching consequences, including the decline or extinction of species, reduced food security, altered nutrient cycles, and disruptions to human societies that depend on healthy ecosystems.

Instances of ecological collapse include the collapse of fisheries due to overfishing, the degradation of coral reefs due to climate change and pollution, and the deforestation-driven collapse of ecosystems in some tropical regions.

Relationship between Carrying Capacity and Ecological Collapse

Balancing Act: Carrying capacity plays a critical role in preventing ecological collapse. When a population exceeds an ecosystem's carrying capacity, it depletes resources and disrupts ecological processes, making the ecosystem more vulnerable to collapse.

Resilience: Ecosystems with lower population levels relative to their carrying capacity are often more resilient to disturbances. They can better absorb changes in environmental conditions and recover from disturbances.

Sustainability: Maintaining human activities within the carrying capacity of ecosystems is a fundamental principle of sustainability. Sustainable practices aim to ensure that resource use and population sizes do not exceed the ecosystem's ability to support them, thus preventing ecological collapse.

Human Impact and Mitigation

Many instances of ecological collapse are driven by human activities, such as deforestation, pollution, and overharvesting. Recognizing the potential consequences, there is a growing focus on sustainable resource management, habitat restoration, and conservation efforts to mitigate and prevent ecological collapse. Addressing ecological collapse often requires international cooperation, as many environmental challenges, such as climate change and the loss of biodiversity, are global in nature. Multilateral agreements and conservation initiatives aim to protect ecosystems and ensure their sustainability.

So, carrying capacity represents the ecological limits of an ecosystem to support populations, while ecological collapse signifies the breakdown of an ecosystem's structure and function due to various stressors. Understanding and respecting carrying capacity is crucial for preventing ecological collapse, conserving biodiversity, and ensuring the long-term health of ecosystems and the services they provide to both nature and society.

4.4. COMPETITIVE EXCLUSION PRINCIPLE

The Competitive Exclusion Principle is a fundamental concept in ecology that explains the dynamics of species coexistence and competition within ecosystems. Proposed by Russian ecologist Georgii Gause in the 1930s, this principle offers insights into how species with similar ecological niches interact and influence each other's survival and distribution.

Definition

The Competitive Exclusion Principle, also known as Gause's Principle, states that two species that occupy the same ecological niche and compete for the same limited resources cannot coexist in the same habitat over an extended period. One species will eventually outcompete and exclude the other, leading to the elimination of the less competitive species.

Key Components

Ecological Niche: The ecological niche of a species encompasses its specific role within an ecosystem, including its habitat, resource utilization, and interactions with other species. Species with highly similar niches are more likely to engage in intense competition.

Resource Competition: The Competitive Exclusion Principle primarily applies to competition for essential resources such as food, water, nesting sites, and sunlight. Species with overlapping resource requirements are more likely to experience competition.

Limited Resources: The principle assumes that resources are limited and cannot support the unlimited growth of all competing species. In reality, most ecosystems have finite resources, which fuel competitive interactions.

Outcomes of Competitive Exclusion

Competitive Superiority: The more competitive species, often referred to as the "superior competitor," will thrive, reproduce, and expand its population. It will eventually dominate the habitat and reduce the population of the inferior competitor.

Local Extinction: The inferior competitor may face reduced access to resources, increased predation, or stress due to competition, leading to a decline in its population. In some cases, it may become locally extinct within the habitat.

Coexistence Mechanisms

Resource Partitioning: In some cases, competing species can coexist through resource partitioning, where they divide available resources by using them in slightly different ways or at different times. This reduces direct competition.

Spatial Heterogeneity: Coexistence can also occur when habitats within an ecosystem vary in resource availability or suitability. This can create refuges or areas where competing species can persist without directly interacting.

Real-World Examples

Darwin's Finches: On the Galápagos Islands, different species of finches have specialized beaks for various types of food, reducing competition for resources and allowing coexistence.

Anolis Lizards: Anolis lizards in the Caribbean exhibit variations in limb length and perch height, reducing competition for perches and food resources.

Significance and Applications

Understanding Species Interactions: The Competitive Exclusion Principle helps ecologists understand how species interactions shape biodiversity, community composition, and the distribution of species in ecosystems.

Conservation and Management: Knowledge of competitive exclusion can inform conservation efforts by identifying species vulnerable to competition and guiding strategies to protect or reintroduce them.

Therefore, the Competitive Exclusion Principle is a foundational concept in ecology that highlights the intense competition for limited resources between species occupying the same ecological niche. While it predicts that one species may ultimately exclude another, real-world examples demonstrate that coexistence mechanisms can allow similar species to share habitats through resource partitioning and spatial heterogeneity.

4.5. ECOLOGICAL YIELD: BALANCING RESOURCE USE AND SUSTAINABILITY

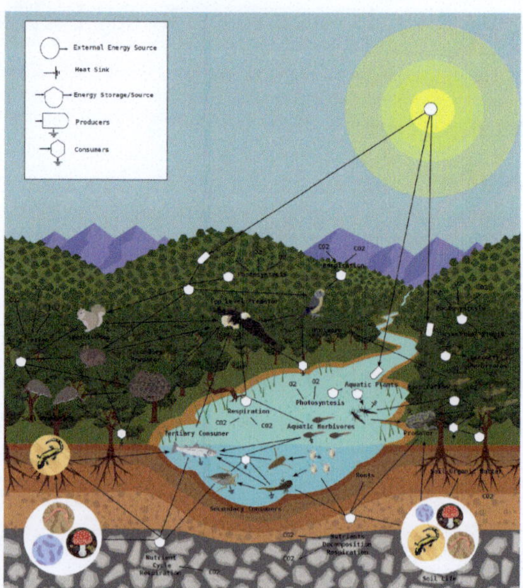

Figure 25. Ecological Yield

(Source: By Original: Thompsma Vector: Pixelsquid – This file was derived from: FoodWeb.jpg: by Thompsma, CC0, https://commons.wikimedia.org/w/index.php?curid=102527971, Creative Commons License)

Ecological yield is a concept in ecology and resource management that relates to the sustainable utilization of natural resources within ecosystems. It focuses on maintaining a balance between the rate at which resources are extracted or harvested and the ecosystem's capacity to regenerate or replenish those resources. The goal of considering ecological yield is to ensure that resource use remains within the limits of what the ecosystem can support over the long term.

Definition

Ecological yield refers to the maximum amount of natural resources that can be harvested, extracted, or used from an ecosystem without depleting those resources or causing long-term harm to the ecosystem's health and functionality.

Components and Considerations

Renewable Resources: Ecological yield primarily applies to renewable resources, which are natural resources that can be regenerated, such as fish stocks, forests, agricultural crops, and groundwater.

Sustainability: The central principle of ecological yield is sustainability. It involves ensuring that resource extraction rates do not exceed the ecosystem's capacity to regenerate those resources naturally.

Rate of Harvest: Ecological yield considers the rate at which resources are harvested or extracted, taking into account factors like population growth, reproduction rates, and the time required for resources to replenish.

Ecosystem Health: Assessing ecological yield involves monitoring the health and resilience of ecosystems to detect any signs of overexploitation, habitat degradation, or disruptions to ecological processes.

Harvesting Practices

Sustainable Practices: Sustainable harvesting practices aim to maintain resource levels within ecological yield limits. These practices may include regulating harvest seasons, setting catch or extraction quotas, and implementing size or age restrictions.

Consequences of Overharvesting: Overharvesting, where resource extraction rates exceed ecological yield, can lead to resource depletion, habitat destruction, and disruptions to ecosystems. This can have negative consequences for both natural ecosystems and human communities that depend on these resources.

Examples

Fisheries Management: Many countries and organizations have adopted fisheries management practices that consider ecological yield to ensure sustainable fishing. This includes setting catch limits based on the estimated reproductive capacity of fish populations.

Forest Management: Sustainable forest management involves selective logging practices that do not exceed the forest's capacity to regenerate timber. It also considers the ecological impacts on forest ecosystems and biodiversity.

Agriculture: Sustainable agriculture seeks to maintain soil fertility and prevent land degradation by using practices that do not deplete soil nutrients or harm the long-term productivity of farmland.

Ecosystem Services

Adhering to ecological yield principles helps preserve essential ecosystem services, such as clean water, pollination, and climate regulation, which benefit both natural ecosystems and human well-being.

Global and Local Considerations

Ecological yield has global implications as overexploitation of resources in one region can affect ecosystems and communities in other parts of the world through interconnected ecological systems and international trade.

Challenges and Trade-offs

Balancing Economic and Environmental Goals: Achieving ecological yield can be challenging, as it often requires balancing economic interests with environmental conservation. Striking the right balance may involve making difficult choices to prioritize sustainability over short-term gains.

Hence, ecological yield is a critical concept for ensuring the sustainable use of renewable resources while preserving the health and functionality of ecosystems. It underlines the importance of managing resource extraction rates within the ecosystem's capacity to regenerate those resources, helping to safeguard the long-term well-being of both natural ecosystems and human societies.

4.6. AUTOTROPH: THE SELF-SUSTAINING PRODUCERS OF THE ECOSYSTEM

Autotrophs are a fundamental group of organisms in ecology that play a vital role in ecosystems by producing their own organic compounds and energy through photosynthesis or chemosynthesis. These self-sustaining organisms form the base of food chains and provide energy and nutrients to all other trophic levels within ecosystems.

Definition

Autotrophs, also known as primary producers, are organisms capable of synthesizing organic compounds (e.g., sugars, starches) from inorganic sources of carbon (usually carbon dioxide) and energy. They do not rely on consuming other organisms for nutrition.

Types of Autotrophs

Photosynthetic Autotrophs: These autotrophs use sunlight as their primary energy source for photosynthesis. They include plants, algae, and certain bacteria (e.g., cyanobacteria or blue-green algae).

Chemosynthetic Autotrophs: Some autotrophs, often found in extreme environments like deep-sea hydrothermal vents and certain underground ecosystems, use chemical compounds as their energy source for chemosynthesis. Examples include certain bacteria and archaea.

Process of Photosynthesis

In photosynthetic autotrophs, the process of photosynthesis involves capturing sunlight through chlorophyll and other pigments to convert carbon dioxide and water into glucose and oxygen. This process releases oxygen into the atmosphere and provides a source of organic matter.

The simplified chemical equation for photosynthesis is 6CO2 + 6H2O + light energy → C6H12O6 + 6O2, where carbon dioxide and water are converted into glucose and oxygen in the presence of sunlight.

Role in Ecosystems

Primary Producers: Autotrophs occupy the first trophic level in food chains and food webs. They serve as the primary source of energy and organic matter for all other organisms in an ecosystem.

Energy Flow: Energy captured by autotrophs during photosynthesis is transferred through the ecosystem as herbivores (primary consumers) consume autotrophs, and then carnivores (secondary and tertiary consumers) consume herbivores. This energy flow ultimately supports all life forms in the ecosystem.

Environmental Impact

Oxygen Production: Photosynthetic autotrophs release oxygen into the atmosphere during photosynthesis, playing a crucial role in maintaining atmospheric oxygen levels, which are essential for aerobic respiration in animals.

Carbon Fixation: Autotrophs also play a vital role in carbon fixation by removing carbon dioxide from the atmosphere and incorporating it into organic compounds. This helps regulate global carbon cycles and mitigate climate change.

Examples

Terrestrial Autotrophs: Plants, including trees, grasses, and shrubs, are some of the most familiar terrestrial autotrophs.

Aquatic Autotrophs: Algae, both microscopic phytoplankton and larger seaweeds, are essential aquatic autotrophs, contributing significantly to oceanic and freshwater ecosystems.

Autotrophs and Heterotrophs

Autotrophs are distinguished from heterotrophs, which obtain their organic compounds and energy by consuming other organisms. Heterotrophs include animals, fungi, and many types of bacteria. So, autotrophs are the foundation of ecosystems, responsible for producing organic matter and providing energy through photosynthesis or chemosynthesis. Their ability to convert inorganic compounds into organic compounds sustains life on Earth by supporting food chains, regulating atmospheric oxygen and carbon dioxide levels, and contributing to the overall health and functioning of ecosystems.

4.7. BACTERIA: THE DIVERSE AND UBIQUITOUS MICROORGANISMS

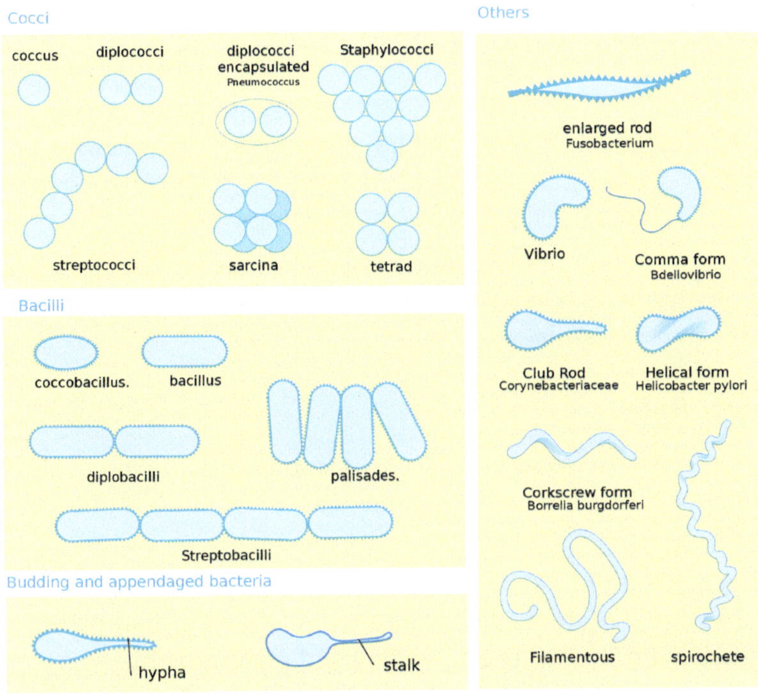

Figure 26. Bacteria

(Source: By Mariana Ruiz Lady of Hats – Own work using: Adobe Illustrator +[1][2][3][4][5][6][7]Medizinische Mikrobiologie, Ernst Wiesmann ED. Thieme (1986), Public Domain, https://commons.wikimedia.org/w/index.php?curid=738916, Creative Commons License)

Bacteria are a diverse group of microorganisms that play a critical role in ecology, human health, industry, and various other fields. These single-celled prokaryotic organisms are among the oldest life forms on Earth and are found in virtually every environment, from deep ocean vents to the human gut. Here, we explore the world of bacteria in greater detail:

Taxonomy and Classification

Prokaryotic Cells: Bacteria are characterized by their prokaryotic cell structure, which means they lack a membrane-bound nucleus and other membrane-bound organelles.

Domain Bacteria: Bacteria belong to the domain Bacteria, one of the three domains of life, with the other two being Archaea and Eukarya. They are distinct from Archaea in terms of genetic, biochemical, and structural differences.

Diversity

Bacteria are incredibly diverse and are classified into numerous phyla and species. Their diversity can be observed in terms of their shapes (e.g., cocci, bacilli, spirilla), metabolic pathways, and ecological roles.

Bacteria are found in extreme environments such as acidic hot springs, deep-sea hydrothermal vents, and polar ice caps, highlighting their adaptability and versatility.

Ecological Roles

Decomposers: Many bacteria serve as decomposers, breaking down dead organic matter into simpler compounds and recycling nutrients back into ecosystems.

Nitrogen Fixation: Some bacteria, like nitrogen-fixing bacteria (e.g., Rhizobium), convert atmospheric nitrogen into a form that plants can use, playing a crucial role in nitrogen cycling and plant growth.

Symbiosis: Bacteria form various symbiotic relationships with other organisms. For example, some live in the guts of animals and aid in digestion (e.g., gut microbiota), while others form nitrogen-fixing root nodules in leguminous plants.

Pathogens: Pathogenic bacteria can cause infectious diseases in humans, animals, and plants. They have been a major driver of human health research and the development of antibiotics.

Industrial and Biotechnological Applications

Fermentation: Bacteria are used in various fermentation processes to produce foods and beverages such as yogurt, cheese, beer, and sauerkraut.

Bioremediation: Some bacteria can break down pollutants and contaminants, making them valuable for bioremediation efforts to clean up environmental pollution.

Biotechnology: Bacteria are used in genetic engineering and biotechnology applications to produce pharmaceuticals, enzymes, and biofuels.

Human Health

Human Microbiota: The human body hosts a diverse community of bacteria, collectively known as the human microbiota. These bacteria play critical roles in digestion, metabolism, and the immune system.

Infectious Diseases: Pathogenic bacteria can cause a wide range of infectious diseases, including strep throat, tuberculosis, cholera, and urinary tract infections.

Antibiotic Resistance

Emergence of Resistance: Overuse and misuse of antibiotics have led to the emergence of antibiotic-resistant bacteria, posing a significant global health threat.

Research and Mitigation: Ongoing research aims to understand and combat antibiotic resistance through the development of new antibiotics and improved stewardship of existing ones.

Bacterial Genetics

Plasmids: Bacteria can exchange genetic material, including plasmids, which are small, circular DNA molecules that can carry genes for antibiotic resistance and other traits.

Horizontal Gene Transfer: Bacteria can undergo horizontal gene transfer, allowing them to acquire new genetic material from other bacteria, archaea, or viruses.

So, bacteria are a diverse group of microorganisms with a wide range of ecological roles and applications. Their adaptability, genetic diversity, and ubiquity in various environments make them a fascinating and essential focus of scientific research and exploration.

4.8. BIO-INVADERS: UNDERSTANDING INVASIVE SPECIES AND THEIR ECOLOGICAL IMPACT

Bioinvaders, often referred to as invasive species, are organisms that have been introduced into new environments where they are not native and have the potential to cause harm to ecosystems, economies, and human health. These non-native species can disrupt local ecosystems, outcompete native species, and have wide-ranging ecological and socio-economic impacts.

Figure 27. Bioinvaders

(Source: https://en.wikipedia.org/wiki/File:Beaver_dam_in_Tierra_del_Fuego.jpg, Creative Commons License)

Definition

Bioinvaders, or invasive species, are plants, animals, fungi, or microorganisms that have been transported by human activities to regions outside their natural habitat or range. Once established, they can proliferate and negatively impact their new environment.

Characteristics and Impact

Rapid Reproduction: Invasive species often reproduce quickly and prolifically, leading to rapid population growth and competitive advantages over native species.

Competitive Advantage: Invasive species can outcompete native species for resources like food, water, and habitat, leading to declines or extinctions of native species.

Ecological Disruption: The introduction of invasive species can disrupt ecological processes, such as nutrient cycling and pollination, which are critical for ecosystem health.

Habitat Alteration: Some invasive species alter their new habitats physically by changing soil composition or the structure of ecosystems, further impacting native species.

Economic and Health Impact: Invasive species can have significant economic consequences by damaging crops, forests, and infrastructure.

Some invasive species also pose health risks to humans and animals by spreading diseases or causing allergic reactions.

Common Examples

Zebra Mussels: Zebra mussels, native to Eastern Europe, have invaded water bodies in North America and Europe, where they disrupt ecosystems by filtering out plankton and altering water quality.

Asian Carp: Asian carp, introduced to North America, threaten native fish populations by outcompeting them for food and habitat.

Kudzu: Kudzu, native to Asia, has become an invasive plant in the southeastern United States, smothering native vegetation.

Cane Toad: The cane toad, introduced to Australia, has had devastating effects on native wildlife due to its toxic skin secretions.

Pathways of Introduction

Human Activities: Invasive species are often introduced unintentionally through human activities like trade, travel, and agriculture. Ballast water from ships, the pet trade, and imported plants and animals are common pathways.

Climate Change: Climate change can alter the range and distribution of species, making some areas more susceptible to invasive species.

Management and Control

Prevention: Preventing the introduction of invasive species through regulations, inspections, and public awareness campaigns is a key strategy.

Early Detection: Early detection and rapid response programs aim to identify and control invasive species before they become established.

Control Methods: Various control methods, including chemical treatments, biological control (using natural predators or pathogens), and physical removal, are used to manage invasive species.

Global Collaboration

Many countries collaborate on efforts to prevent and manage invasive species, recognizing the global nature of the problem.

Conservation Implications

Biodiversity Conservation: Invasive species pose a significant threat to biodiversity and can exacerbate the already critical issue of species extinction.

Restoration: Restoration efforts often involve the removal of invasive species to restore native ecosystems.

In fact, Bioinvaders, or invasive species, are non-native organisms that can have severe ecological, economic, and health impacts when introduced into new environments. Effective management and prevention efforts are essential to mitigate the negative consequences of invasive species and protect native ecosystems and biodiversity.

4.9. BIOMASS/BIOENERGY: HARNESSING RENEWABLE RESOURCES FOR SUSTAINABLE ENERGY

Biomass and bioenergy represent a renewable and environmentally friendly source of energy derived from organic materials. These resources play a significant role in reducing greenhouse gas emissions, promoting sustainability, and diversifying energy sources. Here, we delve into the concepts of biomass and bioenergy in greater detail:

Biomass

Biomass refers to any organic material derived from living or recently living organisms, such as plants, trees, crop residues, and organic waste. It can include wood, agricultural residues, dedicated energy crops, and even algae.

Biomass is considered renewable because it relies on the continuous growth and replenishment of organic materials, primarily through photosynthesis.

Biomass can take various forms, including solid biomass (wood, crop residues), liquid biomass (biofuels like ethanol and biodiesel), and gaseous biomass (biogas and landfill gas).

Bioenergy

Bioenergy refers to the energy derived from the conversion of biomass into useful forms of energy, such as electricity, heat, or biofuels. It encompasses a range of technologies and processes for harnessing energy from organic materials.

Bioenergy can be categorized into several types

Bioelectricity: Generated through the combustion of biomass to produce steam and drive turbines that generate electricity.

Biofuels: Liquid or gaseous fuels produced from biomass, including ethanol, biodiesel, and biogas. They can be used as substitutes for fossil fuels in transportation and heating.

Biogas: Methane-rich gas produced through the anaerobic digestion of organic waste materials, including sewage, agricultural residues, and food waste.

Bioheat: Heat energy obtained from the combustion of biomass, used for space heating, industrial processes, and district heating systems.

Sustainable Aspects

Carbon Neutrality: Biomass and bioenergy are considered carbon-neutral or nearly carbon-neutral when properly managed. This means that the carbon dioxide released during combustion is balanced by the carbon dioxide absorbed by the plants during growth. As such, bioenergy can help mitigate climate change by reducing net greenhouse gas emissions.

Reduced Dependence on Fossil Fuels: The use of biomass and biofuels reduces reliance on fossil fuels, helping to diversify energy sources and enhance energy security.

Waste Utilization: Bioenergy often utilizes organic waste materials, reducing the need for landfill disposal and contributing to waste management and resource efficiency.

Applications

Transportation: Biofuels like ethanol and biodiesel can be used as substitutes for gasoline and diesel fuels in vehicles.

Electricity Generation: Biomass can be burned in power plants to produce electricity. Combined heat and power (CHP) systems can also generate both electricity and heat from biomass.

Heating and Cooling: Biomass can be used for residential and industrial heating, as well as for cooling through absorption cooling systems.

Cooking and Heating in Developing Regions: In many developing regions, biomass, such as wood and crop residues, is the primary source of energy for cooking and heating.

Challenges and Considerations

Sustainability Practices: To ensure the sustainability of bioenergy, responsible practices are essential, including sustainable forestry, efficient crop management, and land-use planning.

Competition with Food Production: The production of biofuels using food crops (e.g., corn for ethanol) can raise concerns about competition for agricultural land and food security.

Land Use and Biodiversity: Expanding bioenergy production can lead to changes in land use, potentially affecting biodiversity and ecosystems.

Technology and Infrastructure: Investment in bioenergy technologies and infrastructure is necessary for widespread adoption.

Future Prospects

Advanced Bioenergy Technologies: Ongoing research focuses on developing advanced bioenergy technologies, such as second-generation biofuels (e.g., cellulosic ethanol), which use non-food plant materials.

Integrated Approaches: Integrated energy systems that combine various renewable sources, including wind, solar, and bioenergy, are being explored to create resilient and sustainable energy networks.

Thus, biomass and bioenergy represent versatile and renewable sources of energy derived from organic materials. Their utilization can contribute to reducing greenhouse gas emissions, enhancing energy security, and promoting sustainable practices in the transition to a more sustainable and environmentally friendly energy future.

4.10. BIOTIC MATERIAL: UNDERSTANDING ORGANIC COMPONENTS IN ECOSYSTEMS

Biotic material refers to the organic components present in ecosystems that are derived from living organisms. These materials play critical roles in ecological processes, nutrient cycling, and the overall functioning of ecosystems. Here, we delve into the concept of biotic material in greater detail:

Definition and Composition

Biotic material encompasses any organic substance or material that originates from living organisms within an ecosystem. It includes a wide range of organic matter, such as plant debris, animal remains, microorganisms, and living organisms themselves.

Sources of Biotic Material

Plant Debris: Fallen leaves, branches, twigs, and dead plant material contribute to biotic material on the forest floor. These materials are collectively referred to as litter.

Animal Remains: Dead animals, including carcasses, bones, shells, and exoskeletons, contribute to biotic material. These materials can serve as a source of nutrients for scavengers and decomposers.

Microorganisms: Microscopic organisms, such as bacteria and fungi, are essential components of biotic material. They play a crucial role in breaking down organic matter through decomposition processes.

Living Organisms: Living plants, animals, and microorganisms themselves are part of the biotic material within an ecosystem. They represent the living component of the ecosystem.

Functions and Significance

Nutrient Cycling: Biotic material is a key player in nutrient cycling within ecosystems. When plants, animals, or microorganisms die, their organic matter contains essential nutrients like carbon, nitrogen, and phosphorus. Decomposers break down this material, releasing nutrients back into the soil, water, and atmosphere.

Energy Transfer: Biotic material serves as a source of energy for decomposers, detritivores (organisms that consume detritus), and scavengers. It forms the basis of food webs and energy flow within ecosystems.

Habitat and Shelter: Biotic material, such as fallen logs or leaf litter, can provide habitat and shelter for various organisms, including insects, fungi, and small mammals.

Carbon Sequestration: In forests and wetlands, biotic material can sequester carbon for extended periods, contributing to carbon storage and climate change mitigation.

Decomposition

Decomposition is the process by which biotic material is broken down into simpler compounds by decomposers, such as bacteria and fungi. During decomposition, organic matter is transformed into humus, a stable form of organic material that enriches the soil and improves its structure.

Decomposition rates are influenced by environmental factors, such as temperature, moisture, and nutrient availability, as well as the type and quality of the organic material.

Human Impact

Land Use Changes: Human activities, such as deforestation, urbanization, and agriculture, can alter the quantity and quality of biotic material in ecosystems.

Waste Management: Human waste, such as sewage and landfill waste, represents a significant source of biotic material that requires proper management to prevent pollution and environmental degradation.

Conservation and Management

Biodiversity Conservation: Recognizing the importance of biotic material in supporting biodiversity, conservation efforts often include the preservation of natural ecosystems and their organic components.

Sustainable Land Use: Sustainable land use practices aim to minimize the disruption of biotic material in ecosystems, protecting both nutrient cycling and habitat availability.

In summarymary, biotic material encompasses the organic components of ecosystems that originate from living organisms. It plays a vital role in nutrient cycling, energy transfer, and the overall functioning of ecosystems. Understanding the dynamics of biotic material is crucial for ecological conservation and sustainable land management.

4.11. CARBON CYCLE

The carbon cycle is a crucial biogeochemical process that describes the movement of carbon, in various forms, through Earth's atmosphere, oceans, land, and living organisms. It is a fundamental cycle that sustains life on our planet, influences climate patterns, and plays a central role in environmental and ecological processes. Let's explore the carbon cycle in detail:

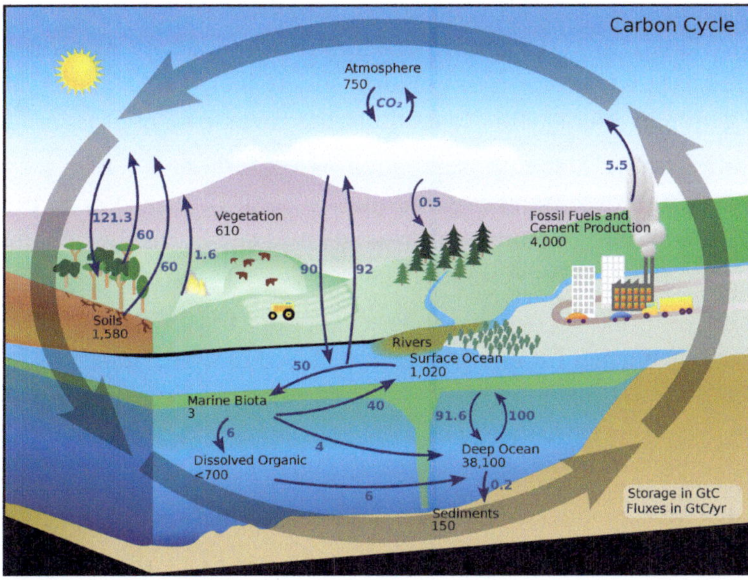

Figure 28. Carbon Cycle

(Source: By Original: Kevin Saff Vector: FischX – Own work based on: Carbon cycle-cute diagram.jpeg (i.e., https://earthobservatory.nasa.gov/features/CarbonCycle), Public Domain, https://commons.wikimedia.org/w/index.php?curid=4134181, Creative Commons License)

Carbon Forms and Reservoirs

Carbon Forms: Carbon exists in various forms in the environment, including carbon dioxide (CO2) in the atmosphere, dissolved carbon compounds in oceans, organic matter in living organisms, and carbon stored in rocks and fossil fuels.

Carbon Reservoirs: These are the major storage locations of carbon on Earth. The primary reservoirs include the atmosphere (as CO2), terrestrial ecosystems (plants, soils, and animals), the oceans, and the Earth's lithosphere (rocks and fossil fuels).

Carbon Cycling Processes

Photosynthesis: This biological process, primarily conducted by plants and algae, converts atmospheric CO_2 into organic carbon compounds, mainly glucose and cellulose, using sunlight and water. During this process, carbon is transferred from the atmosphere into living organisms, forming the foundation of terrestrial food webs. Respiration: All living organisms, including plants, animals, and microorganisms, undergo respiration. This process releases CO_2 back into the atmosphere as organic matter is broken down to generate energy. It is essentially the reverse of photosynthesis.

Decomposition: Decomposers, such as bacteria and fungi, play a vital role in breaking down dead organic matter, returning carbon to the soil or atmosphere as CO_2. Decomposition is a key step in the carbon cycle and contributes to nutrient recycling.

Combustion: When organic matter, such as fossil fuels (coal, oil, and natural gas), is burned for energy, carbon stored in these materials is rapidly released into the atmosphere as CO_2. This process is a significant driver of anthropogenic (human-caused) carbon emissions and contributes to global climate change.

Ocean Uptake: Oceans act as a major carbon sink by absorbing CO_2 from the atmosphere. This absorption process, called oceanic carbon uptake, helps regulate atmospheric CO_2 levels but can lead to ocean acidification, which has adverse effects on marine ecosystems.

Weathering and Erosion: Over long geological time scales, weathering of rocks and erosion processes release carbon from the Earth's lithosphere into surface waters and, eventually, the oceans.

Sedimentation: Carbon in the form of organic material can settle to the ocean floor, forming sedimentary layers over millions of years. Over geologic time, these sediments can become rocks like limestone and shale, effectively storing carbon for extended periods.

Human Impact and Climate Change

Anthropogenic Carbon Emissions: The burning of fossil fuels, deforestation, and other human activities have significantly increased atmospheric CO_2 levels. This excess CO_2 is a primary driver of global warming and climate change.

Feedback Loops: Rising temperatures can lead to positive feedback loops in the carbon cycle, such as permafrost thawing and increased microbial decomposition, which release additional carbon into the atmosphere.

Importance of the Carbon Cycle

Climate Regulation: The carbon cycle plays a central role in regulating Earth's climate by controlling the concentration of greenhouse gases in the atmosphere, primarily CO_2. Ecosystem Functioning: It supports the growth and functioning of ecosystems, influencing plant growth, soil fertility, and nutrient cycling.

Human Survival: Carbon cycling provides essential services like food production, clean water, and air quality, directly impacting human well-being.

Mitigation and Management

Carbon Sequestration: Efforts to mitigate climate change involve carbon sequestration, which aims to capture and store carbon, such as reforestation, afforestation, and carbon capture and storage (CCS) technologies.

Reducing Emissions: Decreasing carbon emissions through sustainable land management, renewable energy adoption, and energy efficiency measures is crucial for mitigating climate change.

In a nutshell, the carbon cycle is a complex and interconnected system that regulates carbon flows between Earth's major reservoirs. Understanding this cycle is essential for addressing climate change, preserving ecosystems, and ensuring the sustainability of life on our planet.

4.12. CLIMATE: UNDERSTANDING EARTH'S LONG-TERM WEATHER PATTERNS

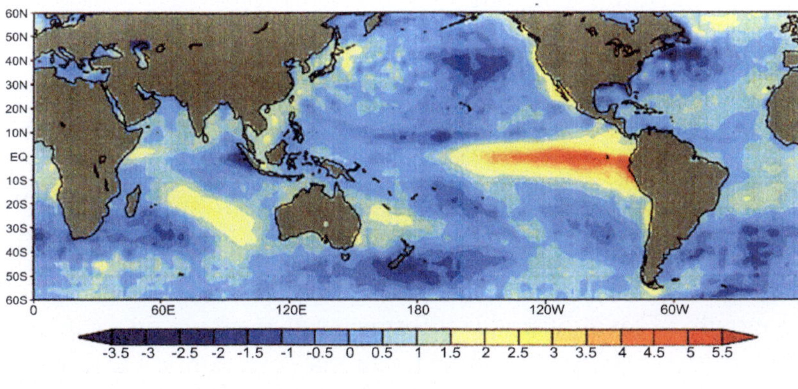

Figure 29. **Climate and Weather Patterns**

(Source: By NOAA. – National Centers for Environmental Prediction, US., Public Domain, https://commons.wikimedia.org/w/index.php?curid=37734223, Creative Commons License)

Climate refers to the long-term patterns of temperature, precipitation, humidity, wind, and other atmospheric conditions in a particular region or on Earth as a whole. Unlike weather, which describes short-term atmospheric conditions, climate provides a broader perspective on the average weather patterns and variations over extended periods. Here, we delve into the concept of climate in greater detail:

Climate vs. Weather

Climate: Climate encompasses the average and long-term weather patterns and conditions in a given region or on Earth as a whole. It is characterized by parameters such as temperature, precipitation, wind patterns, and seasonality over extended periods, typically spanning decades to millions of years.

Weather: Weather refers to the short-term atmospheric conditions that can change from day to day or even within hours. It includes factors like temperature, humidity, precipitation, cloud cover, and wind speed.

Climate Classification

Climate Zones: Earth's climate is divided into different zones based on temperature and precipitation patterns. The Köppen climate classification system is one of the most widely used, categorizing climates into types such as tropical, arid, temperate, and polar.

Factors Influencing Climate

Solar Radiation: The distribution of solar energy from the sun is a primary driver of climate. Regions closer to the equator receive more direct sunlight and tend to be warmer, while polar regions receive less direct sunlight and are colder.

Ocean Currents: Ocean currents, such as the Gulf Stream, can influence the climate of coastal regions by transporting warm or cold water, affecting temperature and precipitation patterns.

Topography: Elevation and landforms, such as mountains, can influence local climate by blocking or channeling air masses, resulting in rain shadows and temperature variations.

Atmospheric Circulation: The movement of air masses, driven by temperature differences and the Coriolis effect, influences wind patterns and climate.

Climate Change

Natural Climate Variability: Earth's climate has naturally varied over geological time scales due to factors like volcanic eruptions, changes in solar radiation, and Earth's orbital variations.

Anthropogenic Climate Change: Human activities, primarily the burning of fossil fuels and deforestation, have significantly increased greenhouse gas concentrations in the atmosphere, leading to global warming and anthropogenic climate change. This has led to more extreme weather events, rising sea levels, and shifts in climate patterns.

Climate Models: Scientists use climate models to simulate past, present, and future climate conditions, helping predict the impacts of climate change and inform policy decisions.

Impact of Climate on Ecosystems and Society

Ecosystems: Climate directly affects ecosystems by determining the types of plants and animals that can thrive in a region, impacting biodiversity, food availability, and habitat suitability.

Agriculture: Climate influences crop growth, water availability, and pest patterns, significantly affecting agricultural practices and food security.

Water Resources: Climate changes can impact the distribution of freshwater resources, leading to water scarcity and competition for water.

Human Settlements: Coastal cities are vulnerable to sea-level rise, and extreme weather events can disrupt infrastructure and human settlements.

Climate Mitigation and Adaptation

Mitigation: Climate mitigation involves efforts to reduce greenhouse gas emissions through actions like transitioning to renewable energy sources, improving energy efficiency, and afforestation.

Adaptation: Climate adaptation strategies focus on preparing for and responding to the impacts of climate change, including building resilient infrastructure, developing drought-resistant crops, and managing water resources effectively.

International Efforts

Paris Agreement: The Paris Agreement is a global accord aimed at addressing climate change by limiting global warming to well below 2 degrees Celsius above pre-industrial levels. It encourages countries to take actions to reduce emissions and enhance climate resilience.

Therefore, climate represents the long-term patterns and averages of weather conditions in a region or on Earth as a whole. It is a dynamic system influenced by various natural and human-induced factors, with significant implications for ecosystems, society, and the global environment. Addressing climate change through mitigation and adaptation measures is crucial for a sustainable future.

4.13. ECOLOGICAL SELECTION: ADAPTATION TO ENVIRONMENTAL PRESSURES

Ecological selection, also known as natural selection or environmental selection, is a fundamental concept in ecology and evolutionary biology. It describes the process by which certain traits or characteristics of organisms become more or less common in a population over time due to their impact on an organism's ability to survive and reproduce within a specific environment. Here, we explore the concept of ecological selection in greater detail:

Natural Selection and Environmental Pressures

Ecological selection is a subset of natural selection that focuses on how environmental pressures, such as competition for resources, predation, climate, and habitat conditions, influence the survival and reproduction of individuals within a population.

Variation in Traits: Within any population, individuals exhibit variations in traits or characteristics, such as size, coloration, behavior, and physiological adaptations.

Differential Fitness: Some traits provide individuals with a better chance of surviving and reproducing in their specific environment. These individuals are said to have higher fitness, as they contribute more offspring to the next generation.

Heritability: Traits that confer advantages in survival and reproduction tend to be passed on to the next generation through genetic inheritance, leading to changes in the population's genetic makeup over time.

Mechanisms of Ecological Selection

Adaptation: Ecological selection is the driving force behind adaptation, where species evolve traits that enhance their ability to thrive in their particular ecological niche.

Directional Selection: In directional selection, a particular trait becomes more common in a population due to the advantage it provides in a changing environment. For example, longer beaks in birds may evolve in response to changes in available food sources.

Stabilizing Selection: Stabilizing selection favors intermediate traits while selecting against extreme variations. This process helps maintain the stability of certain traits within a population.

Disruptive Selection: Disruptive selection favors extreme variations of a trait over intermediate forms. This can lead to the development of multiple distinct phenotypes within a population.

Examples of Ecological Selection

Peppered Moths: The classic example of ecological selection involves peppered moths in industrialized regions. Dark-colored moths became more common in areas with pollution due to their increased ability to blend with soot-covered trees, offering protection from predators.

Darwin's Finches: Charles Darwin's observations of finches on the Galápagos Islands demonstrated ecological selection. Different beak shapes evolved in response to varying food sources on different islands.

Antibiotic Resistance in Bacteria: Bacteria can develop antibiotic resistance through ecological selection. Individuals with mutations conferring resistance have a survival advantage in the presence of antibiotics and are more likely to reproduce.

Human Impact and Conservation

Human Activities: Human activities, such as habitat destruction, pollution, and climate change, can exert strong ecological selection pressures on species, leading to changes in their distributions, behaviors, and adaptations.

Conservation Biology: Understanding ecological selection is crucial in conservation biology, as it helps scientists predict how species may respond to changing environments and informs strategies for preserving biodiversity.

Coevolution and Ecological Relationships

Coevolution: Ecological selection often drives coevolutionary processes, where species interact and exert selection pressures on each other. Examples include predator-prey arms races and mutualistic relationships.

Evolution and Speciation

Speciation: Over long periods, ecological selection can contribute to the process of speciation, where populations of a species diverge into distinct species adapted to different ecological niches.

So, ecological selection is a fundamental mechanism of evolution driven by environmental pressures. It shapes the traits and adaptations of organisms, leading to changes in populations over time. Understanding ecological selection is essential for studying biodiversity, conservation, and the processes that underlie the diversity of life on Earth.

4.14. GAIA HYPOTHESIS: EARTH AS A SELF-REGULATING LIVING ORGANISM

The Gaia Hypothesis, proposed by British chemist and inventor James Lovelock in the 1970s, is a provocative and controversial concept that suggests Earth functions as a self-regulating, living organism. Named after Gaia, the Greek goddess of the Earth, this hypothesis posits that life on our planet interacts with its physical and chemical environment to maintain conditions suitable for life. While not universally accepted within the scientific community, the Gaia Hypothesis has sparked discussions and research on Earth's interconnectedness. Here's a detailed exploration of the Gaia Hypothesis:

Overview

Living Earth: The Gaia Hypothesis suggests that Earth, with all its living organisms, ecosystems, and the physical environment, behaves like a single, self-regulating organism. In essence, it views Earth as a living system.

Homeostasis: Central to the hypothesis is the idea of homeostasis, the ability of a system to maintain stable and favorable conditions for life. Lovelock proposed that the Earth's biota (all living organisms) collaborate to regulate and stabilize the planet's environment, including temperature, atmospheric composition, and ocean salinity.

Regulating Feedback Mechanisms: Lovelock proposed that various feedback mechanisms, both positive (amplifying change) and negative (stabilizing change), operate within Earth's systems to maintain equilibrium. These feedbacks can involve the biosphere (living organisms), atmosphere, and lithosphere (Earth's crust).

Key Concepts and Mechanisms:

Daisyworld Model: Lovelock introduced the Daisyworld model as a simplified representation of the Gaia Hypothesis. In this model, two types of daisies with different colors (black and white) inhabit a planet. These daisies can alter the planet's albedo (reflectivity) and, therefore, its temperature. The model demonstrates how life can influence environmental conditions.

Gas Regulation: One of the most controversial aspects of the hypothesis is the idea that life, particularly microbial life, plays a role in regulating the composition of the atmosphere. For example, some microorganisms can influence the levels of greenhouse gases like methane and carbon dioxide.

Albedo Changes: Changes in surface reflectivity, caused by living organisms like forests or ice cover, can affect the absorption of sunlight and, consequently, temperature.

Sulfur and Climate Regulation: Sulphur-producing bacteria in the oceans were proposed as a mechanism by which Earth could regulate its temperature by controlling the production of cloud-forming aerosols.

Controversy and Criticism

Scientific Skepticism: The Gaia Hypothesis has faced criticism and skepticism within the scientific community. Some scientists argue that the Earth's self-regulation is better explained by non-biological, physical, and chemical processes.

Teleological Language: Critics have raised concerns about the teleological language used in describing the hypothesis, implying purpose or intentionality in Earth's behavior.

Modern Perspectives

Gaia Theory: Lovelock's ideas have evolved into what is now known as Gaia theory. While it retains the core concept of Earth as a self-regulating system, it emphasizes that this regulation is not necessarily conscious or intentional.

Scientific Exploration: Contemporary research continues to investigate the interactions between life and the environment and their role in shaping the planet's climate and conditions.

- **Influence and Legacy**

Ecosystem Science: The Gaia Hypothesis has contributed to a broader appreciation of the interconnectedness of Earth's systems and has influenced fields like ecology, biogeochemistry, and Earth system science.
Environmental Ethics: The hypothesis has also sparked discussions about humanity's role as stewards of the planet and the ethical responsibilities associated with environmental conservation.
In summarymary, the Gaia Hypothesis proposes a thought-provoking perspective on the Earth as a self-regulating, living system. While it has generated considerable debate, it has also inspired research and discussions on the intricate relationships between life and the environment.

4.15. NATURAL RESOURCE: THE FOUNDATION OF HUMAN CIVILIZATION

Natural resources are the raw materials and components found in the natural environment that are essential for the sustenance and advancement of human civilization. These resources can be categorized into various types, including renewable and non-renewable resources, and are fundamental to the economic, social, and environmental aspects of human life. Here, we provide a detailed exploration of natural resources:

- **Types of Natural Resources**

Renewable Resources: These are resources that can be naturally replenished over time. Examples include sunlight, wind, water, timber, and biomass.
Non-Renewable Resources: These are finite resources that cannot be quickly replenished, and their supply is limited. Examples include fossil fuels (coal, oil, natural gas), minerals (such as iron and copper), and nuclear fuels (uranium).
Flow Resources: These resources are continuously available and do not deplete with use. Examples include solar energy, wind energy, and tidal energy.
Stock Resources: These resources have a limited supply that can be exhausted if not managed sustainably. Examples include groundwater, minerals, and fossil fuels.

- **Significance and Utilization**

Economic Development: Natural resources are essential for economic growth and development. They serve as inputs for various industries, in-

cluding agriculture, manufacturing, energy production, and construction.
Energy Production: Fossil fuels, such as oil, coal, and natural gas, provide the majority of the world's energy needs. Renewable resources like wind, solar, and hydropower are also crucial for clean energy production.
Food and Agriculture: Arable land, freshwater, and fertile soil are vital for food production. Natural resources like forests and oceans contribute to the global food supply.
Infrastructure and Construction: Minerals, metals, and construction materials are essential for building infrastructure, homes, and transportation systems.
Medicine and Pharmaceuticals: Natural resources provide raw materials for the pharmaceutical and medical industries, including plant-derived medicines and minerals used in medical equipment.
Biodiversity and Ecosystem Services: Biodiversity is a natural resource that provides ecosystem services like pollination, water purification, and climate regulation.

- **Sustainable Management and Conservation**

Resource Depletion: Overexploitation and unsustainable use of natural resources can lead to depletion and environmental degradation. Sustainable management practices are crucial to ensure the availability of resources for future generations.
Protected Areas: Conservation efforts often involve creating protected areas, national parks, and marine reserves to preserve ecosystems and their associated resources.
Sustainable Agriculture: Practices such as organic farming, crop rotation, and agroforestry aim to use natural resources in agriculture more sustainably.

- **Environmental Impact**

Climate Change: The extraction and burning of fossil fuels release greenhouse gases, contributing to global climate change.
Habitat Destruction: Deforestation, mining, and land development can lead to habitat destruction and biodiversity loss.
Pollution: Resource extraction and industrial processes can result in pollution of air, water, and soil, affecting ecosystems and human health.

- **Future Challenges**

Resource Scarcity: As the global population grows and economies expand, there is increasing pressure on natural resources, leading to concerns about resource scarcity.
Transition to Renewable Energy: The transition from fossil fuels to renew-

able energy sources is a significant challenge but essential for mitigating climate change. Sustainable Consumption: Promoting sustainable consumption patterns and reducing waste is crucial for resource conservation.

In summarymary, natural resources are the foundation of human civilization, providing the materials and energy needed for economic, social, and environmental well-being. Sustainable management and conservation efforts are essential to ensure the long-term availability of these resources and to address global challenges like climate change and resource scarcity.

4.16. MONOCULTURE: THE AGRICULTURAL PRACTICE OF CULTIVATING A SINGLE CROP

Monoculture is an agricultural practice in which a single type of crop is cultivated in a large area of land over an extended period. In contrast to polyculture, where multiple crops are grown together, monoculture focuses on the mass production of one specific crop. This practice has both advantages and disadvantages and plays a significant role in modern agriculture. Here, we explore the concept of monoculture in greater detail:

Key Characteristics of Monoculture

Single Crop Species: Monoculture involves the cultivation of a single crop species, such as wheat, rice, corn, or soybeans, across a large agricultural area.

Large-Scale Production: It is typically practiced on a large scale, with extensive fields or plantations dedicated to a single crop.

Consistency: Monoculture aims for consistency and uniformity in crop production, making it easier to manage planting, harvesting, and pest control.

Mechanization: Mechanized farming practices, including the use of tractors, combines, and specialized equipment, are often associated with monoculture for efficiency.

Advantages of Monoculture

Efficient Use of Resources: Monoculture allows for the efficient use of resources such as land, labor, and machinery, as the same crop is grown throughout the entire area.

High Yields: Focusing on a single crop can lead to higher yields per unit of land because the crop is optimized for growth and production.

Simplified Management: Managing a single crop simplifies tasks like planting, fertilization, and pest control, reducing labor and operational costs. Uniformity: Monoculture can result in uniform crops that are easier to harvest and process.

Economies of Scale: Large-scale monoculture operations often benefit from economies of scale, which can lead to lower production costs.

Disadvantages and Concerns

Pest Vulnerability: Monoculture can make crops more vulnerable to pests and diseases, as a single crop provides a consistent food source for specific pests. This can lead to increased pesticide use.

Soil Degradation: Continuous cultivation of the same crop can deplete soil nutrients and reduce soil fertility, leading to decreased crop yields over time.

Biodiversity Loss: Monoculture can have negative impacts on biodiversity, as it reduces the variety of plants and animals in the area.

Environmental Impact: Intensive monoculture agriculture can lead to environmental issues like soil erosion, water pollution from agricultural runoff, and habitat destruction.

Resilience: Monoculture systems may be less resilient to environmental changes, such as climate variability and extreme weather events, which can disrupt crop production.

Alternatives and Sustainable Practices

Crop Rotation: Crop rotation involves alternating the types of crops grown in a particular field to reduce soil depletion and pest pressure.

Agroforestry: Agroforestry combines tree and crop cultivation, enhancing biodiversity and promoting sustainable land use.

Polyculture: Polyculture involves growing multiple crops together, which can improve soil health and reduce pest pressure.

Organic Farming: Organic farming practices often focus on diversity, soil health, and reduced chemical inputs, addressing some of the concerns associated with monoculture.

In fact, monoculture is a common agricultural practice that focuses on the mass production of a single crop. While it offers advantages in terms of efficiency and high yields, it also raises concerns about environmental sustainability, soil health, and biodiversity. As a result, many farmers and researchers are exploring alternative and more sustainable farming practices to address these challenges.

4.17. DECOMPOSITION OF ORGANIC MATTER: NATURE'S RECYCLING PROCESS

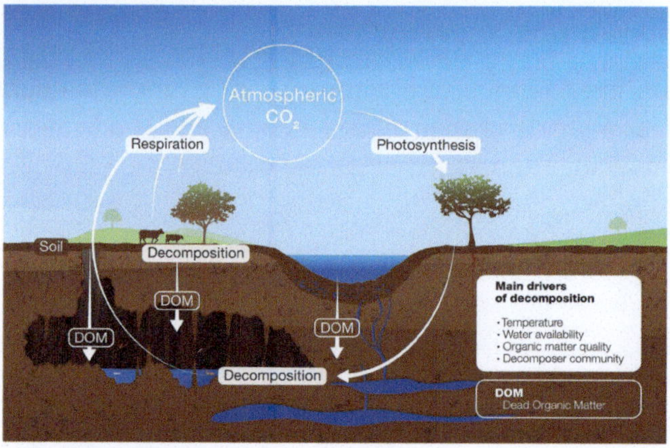

Figure 30. Decomposition of Organic Matter

(Source: Ravn, N. R., Michelsen, A., & Reboleira, A. S. P. S. (2020). Decomposition of Organic Matter in Caves. Front. Ecol. Evol. 8, 554651. doi: 10.3389/fevo.2020.554651, Creative Commons License)

Organic matter decomposition is a vital ecological process in which complex organic compounds, derived from living organisms, are broken down into simpler substances by decomposers like bacteria, fungi, and detritivores. This process plays a fundamental role in nutrient cycling and the recycling of organic materials in ecosystems. Here, we delve into the decomposition of organic matter and its significance:

Organic Matter Decomposition Process:

Organic Substances: Organic matter primarily consists of carbon-based compounds derived from living organisms. This includes dead plant material (leaves, branches), animal remains (carcasses, feces), microorganisms, and other biological material.

Decomposer Community: Decomposers are organisms that specialize in breaking down organic matter. This community includes bacteria, fungi (such as saprophytic mushrooms), earthworms, insects, and other detritivores.

Enzyme Action: Decomposers secrete enzymes that break down complex organic molecules like cellulose, lignin, and proteins into simpler substances. These enzymes catalyze chemical reactions that convert large organic molecules into soluble compounds.

Mineralization: During decomposition, complex organic molecules are converted into inorganic forms, such as carbon dioxide (CO2), water, and various mineral ions. These inorganic forms can be absorbed by plants as nutrients.

Significance of Organic Matter Decomposition:

Nutrient Recycling: Decomposition releases essential nutrients, including carbon, nitrogen, phosphorus, and sulfur, back into the soil. These nutrients become available for plants and microorganisms, completing nutrient cycles.

Soil Formation: Organic matter decomposition contributes to the formation of humus, a stable, dark, and nutrient-rich component of soil. Humus improves soil structure, water retention, and fertility.

Energy Flow: Decomposition is a crucial step in energy flow within ecosystems. It transfers energy from dead organic matter to decomposers and, eventually, to higher trophic levels through consumption.

Waste Management: Decomposers play a critical role in recycling organic waste, including dead plant material, fallen leaves, and animal remains. This reduces litter buildup and helps maintain ecosystem cleanliness.

Factors Influencing Decomposition:

Environmental Conditions: Temperature, moisture, oxygen availability, and pH levels significantly influence decomposition rates. Warmer and moister conditions generally accelerate decomposition.

Chemical Composition: The chemical composition of organic matter, such as the C:N ratio (carbon-to-nitrogen ratio), affects decomposition. Materials with high C:N ratios may decompose more slowly due to nitrogen limitation.

Detritivore Activity: The presence and activity of detritivores like earthworms, insects, and scavengers can accelerate decomposition by physically breaking down organic matter.

Human Impact and Decomposition:

Land Use Changes: Human activities like deforestation, urbanization, and agriculture can alter decomposition rates and soil health.

Waste Management: Proper waste management practices, including composting and recycling, harness decomposition processes to reduce waste and environmental pollution.

Climate Change: Altered precipitation patterns and temperature shifts associated with climate change can impact decomposition rates, potentially affecting nutrient cycling.

So, the decomposition of organic matter is a fundamental ecological process that recycles nutrients, contributes to soil fertility, and sustains ecosystems. Understanding the factors influencing decomposition is crucial for managing land use, waste, and soil health while addressing environmental challenges like climate change and nutrient cycling.

4.18. INORGANIC SUBSTANCE: DETRITUS AND BIO-DEGRADATION

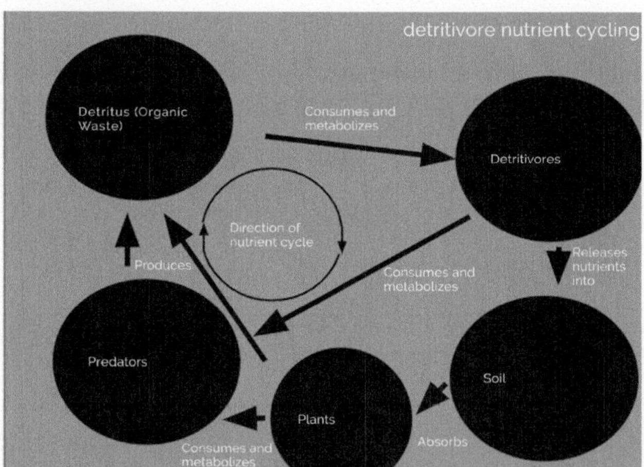

Figure 31. Inorganic Substance: Detritus and Biodegradation

(Source: By Kenneth 28232 – Own work, CC BY-SA 4.0, https://commons.wikimedia.org/w/index.php?curid=130447425, Creative Commons License)

Inorganic substances are chemical compounds or elements that lack carbon-hydrogen (C-H) bonds, which are characteristic of organic compounds. Inorganic substances are fundamental to the physical and chemical processes that shape Earth's geology, chemistry, and biology. They play crucial roles in various natural processes and are essential components of the environment. Here, we explore the concept of inorganic substances, with a focus on detritus and biodegradation:

Detritus: Organic and Inorganic Matter

Detritus refers to a mixture of organic and inorganic matter that consists of small, fragmented, and decaying pieces of dead organisms, debris, and other materials found in ecosystems.

Detritus can include organic materials like leaves, twigs, dead animals, and feces, as well as inorganic materials like minerals, rocks, and soil particles.

Detritus is a critical component of ecosystems, serving as a primary source of energy and nutrients for decomposers and detritivores. It contributes to nutrient cycling, soil formation, and the overall health of ecosystems.

Biodegradation: Breaking Down Organic Compounds

Biodegradation is a natural process by which microorganisms, such as bacteria, fungi, and other decomposers, break down organic compounds into simpler substances. This process occurs in both natural and human-made environments.

Organic Matter: Biodegradation primarily targets organic matter, including complex organic molecules like cellulose, lignin, proteins, and carbohydrates.

Decomposers: Microbial decomposers play a central role in biodegradation. They secrete enzymes that catalyze chemical reactions, breaking down complex organic molecules into smaller, soluble compounds.

Environmental Significance: Biodegradation is essential for recycling nutrients, releasing carbon dioxide (CO2) back into the atmosphere, and maintaining ecosystem health. It also helps manage organic waste and pollutants.

Inorganic Substances in Nature:

Minerals: Minerals are naturally occurring inorganic substances with a specific chemical composition and crystalline structure. They form the Earth's crust and are essential for geological processes.

Water: Water (H2O) is an inorganic compound that is vital for life and various geological processes. It plays a crucial role in weathering rocks, supporting aquatic ecosystems, and facilitating chemical reactions.

Salts: Inorganic salts, such as sodium chloride (table salt) and calcium carbonate, are common inorganic substances found in nature. They contribute to the composition of soils, water bodies, and minerals.

Gases: Inorganic gases like oxygen (O2), nitrogen (N2), and carbon dioxide (CO2) are essential components of Earth's atmosphere and are critical for biological and geological processes.

Human Impact and Inorganic Substances:

Mining: Human activities like mining and mineral extraction can disrupt geological formations and release inorganic substances, potentially leading to environmental pollution and habitat destruction.

Water Quality: Inorganic substances, such as heavy metals and salts, can negatively impact water quality and aquatic ecosystems when released into water bodies as pollutants.

Industrial Processes: Many industrial processes involve the use and release of inorganic substances, affecting air and water quality.

In short, inorganic substances are essential components of Earth's chemistry and play significant roles in geological, biological, and environmental processes. Detritus, composed of both organic and inorganic matter, contributes to ecosystem health, while biodegradation is a crucial process for breaking down organic compounds. Understanding the roles and interactions of inorganic substances in natural systems is vital for environmental management and conservation.

4.19. ECOLOGICAL CRISIS

An ecological crisis refers to a critical, often sudden, and widespread disruption or deterioration of the natural environment and ecosystems, leading to severe negative impacts on life forms, including humans. These crises result from various human activities, natural events, or a combination of both and have far-reaching consequences for biodiversity, ecosystems, and the overall health of the planet. Here, we explore the concept of ecological crises, their causes, consequences, and potential solutions:

Types of Ecological Crises:

Biodiversity Loss: This crisis involves the rapid decline and extinction of plant and animal species due to habitat destruction, pollution, overharvesting, and invasive species.

Climate Change: The alteration of global climate patterns due to the accumulation of greenhouse gases, primarily carbon dioxide (CO2), in the atmosphere is a significant ecological crisis. It leads to rising temperatures, sea-level rise, extreme weather events, and altered ecosystems.

Deforestation: The extensive clearing of forests for agriculture, logging, and urbanization is a crisis that impacts biodiversity, carbon storage, and the water cycle.

Pollution: Ecological crises related to pollution include air pollution, water pollution, and soil contamination, which harm both terrestrial and aquatic ecosystems and pose health risks to humans.

Habitat Loss and Fragmentation: The transformation of natural landscapes into fragmented and isolated habitats through urbanization and infrastructure development disrupts ecosystems and reduces habitat availability for species.

Invasive Species: The introduction of non-native species into new ecosystems can lead to ecological crises, as these invaders outcompete native species, disrupt food webs, and alter ecosystem dynamics.

Overexploitation: Overharvesting of natural resources, such as fish stocks, forests, and minerals, can deplete these resources and lead to ecological imbalances.

Causes of Ecological Crises:

Human Activities: Most ecological crises have human activities at their root, including industrialization, urbanization, agriculture, deforestation, and the burning of fossil fuels.

Population Growth: The increasing human population places pressure on natural resources and ecosystems, leading to habitat destruction and overconsumption.

Consumerism: The modern consumer-driven economy encourages resource-intensive lifestyles and the production of disposable goods, contributing to waste and pollution.

Climate Change: The burning of fossil fuels for energy production and transportation is a major contributor to climate change, which exacerbates many ecological crises.

Globalization: The global movement of goods, people, and information can facilitate the spread of invasive species and pollutants.

Consequences of Ecological Crises:

Loss of Biodiversity: Ecological crises often result in species extinctions and reduced genetic diversity, diminishing the resilience of ecosystems.

Altered Ecosystem Services: Ecosystems affected by crises may provide fewer services like clean water, pollination, and climate regulation, impacting human well-being.

Economic Impact: Ecological crises can have severe economic consequences, including crop failures, property damage from extreme weather events, and healthcare costs from pollution-related illnesses.

Social and Humanitarian Issues: Ecological crises can lead to displacement, food and water scarcity, and social conflicts, particularly in vulnerable communities.

Mitigation and Solutions:

Conservation and Restoration: Protecting and restoring natural habitats, implementing conservation measures, and creating protected areas are crucial for mitigating biodiversity loss and habitat degradation.

Sustainable Practices: Transitioning to sustainable agriculture, forestry, fisheries, and energy production can help reduce the impact of human activities on ecosystems.

Policy and Regulation: Governments and international bodies play a vital role in enacting policies and regulations to address ecological crises, including emissions reduction targets, pollution control, and conservation laws.

Public Awareness and Education: Raising awareness about ecological crises and their consequences can lead to changes in individual and collective behavior, encouraging sustainable practices.

International Cooperation: Addressing global ecological crises often requires international collaboration to develop and implement solutions. In summarymary, ecological crises encompass a range of challenges, from biodiversity loss and climate change to pollution and habitat destruction. These crises are primarily driven by human activities and pose significant threats to the environment, biodiversity, and human well-being. Addressing these challenges requires urgent and coordinated efforts at local, national, and international levels to promote sustainability and protect the planet's ecosystems.

4.20. ECOLOGICAL EXTINCTION

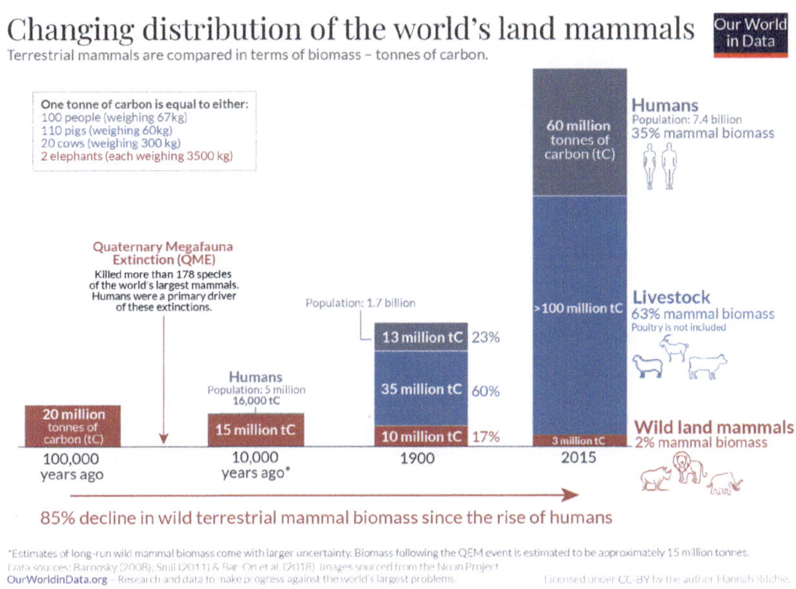

Figure 32. Ecological Extinction

(Source: By Hannah Ritchie, Our World in Data – https://ourworldindata.org/mammals, CC BY 4.0, https://commons.wikimedia.org/w/index.php?curid=120802173, Creative Commons License)

Ecological extinction, also known as functional extinction or ecological collapse, occurs when a species within an ecosystem experiences such a significant decline in numbers that it can no longer fulfill its ecological roles and functions. While the species may still exist in small numbers, its population has dwindled to a point where it no longer has a substantial impact on its ecosystem. Ecological extinction is a critical concept in conservation biology and ecology, as it highlights the intricate relationships between species and the fragility of ecosystems. Here, we explore the concept of ecological extinction in detail:

Key Characteristics of Ecological Extinction

Functional Loss: Ecological extinction emphasizes the loss of a species' ecological function within an ecosystem, rather than its complete disappearance. Even if a few individuals of the species remain, they may not be sufficient to carry out their ecological roles.

Impact on Ecosystem: When a species undergoes ecological extinction, it can disrupt the balance of the ecosystem. This can lead to cascading effects that affect other species, altering predator-prey relationships, nutrient cycling, and overall ecosystem health.

Potential for Recovery: Ecological extinction does not necessarily mean the species can never recover. With conservation efforts and favorable conditions, it may be possible to restore a species' ecological role within an ecosystem.

Causes of Ecological Extinction

Habitat Loss and Fragmentation: Destruction of natural habitats, such as deforestation and urbanization, can lead to the isolation of species populations, making them more vulnerable to ecological extinction.

Overexploitation: Overharvesting of resources, including fishing, hunting, and logging, can deplete populations to the point of ecological extinction.

Invasive Species: The introduction of invasive species can outcompete native species, disrupt food webs, and lead to ecological extinction of vulnerable native species.

Pollution: Pollution, including chemical contaminants and habitat degradation, can harm species and their habitats, driving them toward ecological extinction.

Climate Change: Alterations in temperature and precipitation patterns due to climate change can affect species' distribution and ability to carry out their ecological functions.

Consequences of Ecological Extinction

Cascading Effects: The loss of a species' ecological function can trigger cascading effects throughout the ecosystem. For example, the decline of a top predator can lead to an increase in prey species, affecting vegetation and other trophic levels.

Altered Ecosystem Services: Ecological extinction can result in the loss or alteration of ecosystem services, including pollination, pest control, and nutrient cycling, which can impact human well-being.

Biodiversity Loss: Ecological extinction contributes to the overall loss of biodiversity, which reduces the resilience of ecosystems to environmental changes.

Ecosystem Vulnerability: Ecosystems with multiple cases of ecological extinction may become more vulnerable to additional disturbances and environmental stressors.

Conservation and Mitigation

Habitat Conservation: Protecting and restoring natural habitats is essential for preventing ecological extinction by providing suitable environments for species.

Reducing Overexploitation: Implementing sustainable resource management practices, such as fisheries regulations and wildlife conservation laws, can help prevent overexploitation.

Invasive Species Control: Managing and controlling invasive species can prevent them from causing ecological extinction among native species.

Pollution Reduction: Reducing pollution and adopting sustainable land use practices can help protect species and their habitats.

Climate Change Mitigation: Addressing climate change through emissions reduction and adaptation strategies is crucial for preserving ecosystems and preventing ecological extinction.

In conclusion, ecological extinction highlights the vulnerability of species and ecosystems to human activities and environmental changes. It underscores the importance of conservation efforts and sustainable practices to protect the intricate web of life on Earth and maintain the ecological functions that support all living organisms.

4.21. ECOPHAGY

Ecophagy, a term derived from the Greek words "oikos" (meaning house or environment) and "phagein" (meaning to eat), is a speculative concept that envisions the use of self-replicating nanotechnology, often referred to as "nanobots" or "nanorobots," to consume and transform the Earth's entire biosphere or other natural environments. While this concept is primarily a product of science fiction and remains theoretical, it raises several concerns and ethical considerations. Here, we delve into the concept of ecophagy in greater detail:

Speculative Nature

Ecophagy is a speculative and futuristic concept that falls within the realm of science fiction and speculative fiction. It is not currently a technological reality, and there is no practical evidence or application of ecophagy.

The concept of ecophagy has been explored in science fiction literature and discussions about the potential consequences of advanced nanotechnology.

Hypothetical Mechanisms

Nanotechnology: Ecophagy envisions the use of self-replicating nanobots that could consume organic matter, such as plants, animals, and microorganisms, at a molecular level. These nanobots would break down and transform the Earth's biosphere into different materials or structures.

In speculative scenarios, the purpose of ecophagy is often depicted as a means of resource utilization or transformation, where the Earth's environment is modified to serve specific human or non-human goals.

Concerns and Ethical Considerations

Environmental Destruction: The concept of ecophagy raises profound ethical concerns, primarily related to the potentially catastrophic environmental destruction it could cause. The complete transformation or consumption of the biosphere would likely result in mass extinction and ecological collapse.

Loss of Biodiversity: Ecophagy could lead to the irreversible loss of biodiversity, including plant and animal species, and disrupt the intricate web of ecological relationships that sustain life on Earth.

Ethical Responsibility: The use of such technology would raise significant questions about ethical responsibility, including whether humans have the right to make decisions that could have far-reaching and destructive consequences for the natural world.

Unintended Consequences: The complexity of ecological systems makes it challenging to predict the full range of consequences that ecophagy might have. Unintended outcomes could include the disruption of global climate patterns, loss of ecosystem services, and unforeseen ecological imbalances.

Current State of Nanotechnology

Nanotechnology Today: While nanotechnology has made significant advances in various fields, such as medicine and materials science, the development of self-replicating nanobots capable of ecophagy is currently a topic of science fiction and theoretical speculation. The technical challenges and ethical considerations associated with such technology are immense.

In fact, ecophagy is a speculative concept that envisions the use of advanced nanotechnology to consume and transform natural environments, including the Earth's biosphere. While it is not a current technological reality, it raises significant concerns and ethical questions about the potential consequences of manipulating and altering the natural world on such a vast scale. It serves as a cautionary reminder of the need for responsible and ethical consideration of technological advancements and their potential impacts on the environment and ecosystems.

4.22. ECOLOGICAL NICHE AND NICHE DIFFERENTIATION

Figure 33. Ecological Niche and Niche Differentiation

(Source: ib.bioninja.com.au, Creative Commons License)

An ecological niche is a fundamental concept in ecology that describes the specific functional role and position of a species within its ecosystem. It encompasses the set of environmental conditions and resources that a species requires for survival and reproduction, as well as its interactions with other species. The concept of niche helps ecologists understand how species coexist and how ecosystems maintain biodiversity. Here, we delve into the concept of ecological niche and its importance:

Components of the Ecological Niche

Habitat: The physical location or environment where a species can be found.

Food Source: The specific organisms or resources a species consumes as part of its diet.

Abiotic Conditions: The non-living environmental factors, such as temperature, humidity, and pH, that affect a species' survival and growth.

Reproductive Requirements: The conditions necessary for reproduction, including mating rituals, breeding sites, and reproductive timing.

Interactions: The species' relationships with other organisms, such as predators, prey, competitors, and mutualistic partners.

Significance of Ecological Niches

Resource Partitioning: Ecological niches help explain how multiple species can coexist in the same ecosystem by partitioning resources and minimizing competition. Species with similar niches are more likely to compete directly for resources.

Biodiversity: Niches contribute to biodiversity by allowing multiple species to occupy different niches within an ecosystem, promoting species diversity and ecological stability.

Ecosystem Functioning: The combination of species and their niches determines how an ecosystem functions. Each species plays a unique role in nutrient cycling, energy flow, and ecosystem services.

Conservation: Understanding the niches of endangered or threatened species is essential for effective conservation efforts. Preserving suitable niches is critical for their survival.

Niche Differentiation: Coexistence Strategies

Competitive Exclusion: When two species with very similar niches compete for the same resources, one may eventually outcompete the other, leading to competitive exclusion. Niche differentiation helps prevent this.

Character Displacement: Over time, species in overlapping niches may evolve to reduce competition. This can lead to character displacement, where their traits or resource use patterns diverge.

Sympatric Speciation: In some cases, niche differentiation can lead to sympatric speciation, where species diverge and become reproductively isolated despite living in the same geographical area.

Coexistence Strategies: Species that share similar niches may evolve different strategies to coexist, such as temporal (using resources at different times), spatial (occupying different parts of the habitat), or behavioral (using distinct foraging techniques).

Examples of Ecological Niches

Woodpeckers: These birds have specialized niches for drilling into tree bark to find insects. Different woodpecker species may have varying-sized bills, foraging behaviors, and habitat preferences, allowing them to coexist.

Darwin's Finches: On the Galápagos Islands, different finch species occupy niches based on beak size and shape, allowing them to exploit different food sources, such as seeds, insects, or nectar.

Cacti: Various cactus species have evolved different ecological niches based on their adaptations to specific arid environments. Some cacti specialize in water storage, while others prioritize rapid growth or drought resistance.

So, the ecological niche is a central concept in ecology that defines a species' functional role within its ecosystem. Niche differentiation and coexistence strategies are crucial for maintaining biodiversity and understanding how species interact and adapt to their environments. The study of ecological niches helps elucidate the complex relationships that shape ecosystems and the rich tapestry of life on Earth.

4.23. BIOLOGICAL INTERACTION: NEUTRALISM, AMENSALISM, AND ECOLOGICAL FACILITATION

Biological interactions are the relationships between different species in an ecosystem, and they play a crucial role in shaping ecological communities. These interactions can be categorized into various types, each with its own impact on the participating species. Here, we explore three specific types of biological interactions: neutralism, amensalism, and ecological facilitation.

Neutralism

Neutralism is a type of biological interaction where two species coexist in close proximity, but their interactions have neither a positive nor a negative impact on either species. In neutral interactions, the species do not compete for resources, nor do they benefit from each other. These interactions are often characterized by a lack of noticeable influence on either species.

Examples of Neutralism:

Two Non-Interacting Plant Species: Two plant species growing in different parts of a forest, where they neither compete for resources nor interact in any discernible way.

Fish in Different Ocean Layers: Fish species living in different layers of the ocean (e.g., surface-dwelling fish and deep-sea fish) may have neutral interactions as they occupy different ecological niches.

Amensalism:

Amensalism is a biological interaction where one species is negatively affected (inhibited or harmed), while the other is neither benefited nor harmed. In these interactions, the negative impact is usually due to the release of chemicals or other factors by one species that inhibit the growth or survival of the other species.

Examples of Amensalism:

Allelopathic Plants: Some plants release chemicals into the soil that inhibit the growth of nearby plants, reducing their chances of survival.

Penicillium Mold and Bacteria: Penicillium mold produces the antibiotic penicillin, which inhibits the growth of bacteria without benefiting the mold itself.

Ecological Facilitation:

Ecological facilitation, also known as positive interaction, occurs when one species benefits from the presence of another species, while the second species may also benefit, experience no change, or be unaffected. These interactions often involve species providing shelter, resources, or other ecological advantages to each other.

Examples of Ecological Facilitation:

Nurse Plants and Seedlings: Some plants act as nurse plants by providing shade and shelter to seedlings of other plant species, allowing them to establish and grow in harsh environments.

Burrowing Animals and Other Species: Burrowing animals, like prairie dogs, create underground tunnels that can be used by other species for shelter, such as burrowing owls and reptiles.

Nitrogen-Fixing Bacteria and Legumes: Nitrogen-fixing bacteria form mutualistic relationships with leguminous plants. These bacteria convert atmospheric nitrogen into a form that the plant can use, benefiting both the plant and the bacteria.

In summary, biological interactions, including neutralism, amensalism, and ecological facilitation, play diverse roles in shaping ecosystems. Neutralism describes interactions with no discernible impact, amensalism involves one species negatively affecting another, and ecological facilitation involves species benefiting from each other's presence. Understanding these interactions is essential for comprehending the complexity and dynamics of ecological communities.

4.24. COEXISTENCE THEORY

Coexistence theory is a branch of ecology that seeks to explain how multiple species with similar ecological requirements can coexist within the same habitat or ecosystem. It explores the mechanisms, processes, and conditions that allow species to share resources and persist together over time. Coexistence theory is essential for understanding biodiversity, community structure, and the factors that maintain ecological diversity. Here, we delve into the key concepts and principles of coexistence theory:

Competitive Exclusion Principle

The concept of competitive exclusion was developed by Russian ecologist Georgy Gause in the 1930s. It states that two species competing for the same limited resources cannot coexist indefinitely. One species will eventually outcompete and eliminate the other.

Resource Partitioning: In practice, many species that appear to compete for the same resources have evolved mechanisms to coexist by partitioning resources. This can include differences in foraging behavior, habitat use, or temporal resource utilization.

Niche Differentiation

The ecological niche of a species encompasses its role in the ecosystem, including its habitat, diet, reproductive strategies, and interactions with other species.

When two or more species have overlapping niches, they may compete for the same resources, leading to potential displacement or exclusion.

Niche differentiation refers to species evolving distinct niches to reduce competition.

Coexistence Mechanisms

Predator-Mediated Coexistence: The presence of predators can prevent competitive exclusion by limiting the abundance of dominant competitors, allowing less competitive species to persist.

Disturbance-Mediated Coexistence: Periodic disturbances, such as wildfires or flooding, can create temporary openings in ecosystems, benefiting species adapted to colonize disturbed areas.

Spatial Heterogeneity: Environmental heterogeneity across space can create microhabitats where different species thrive, reducing competition.

Apparent Competition: Two species that do not directly compete for resources may still influence each other through shared predators or pathogens, which can lead to their coexistence.

Theoretical Models

Lotka-Volterra Competition Models: These mathematical models describe the dynamics of competing species in terms of population growth, resource consumption, and competition coefficients. They help predict conditions for coexistence or exclusion.

Resource-Ratio Hypothesis: This hypothesis suggests that the coexistence of species depends on the ratio of the critical resources they require. When the resource ratios differ among species, coexistence is more likely.

Alternative Stable States: Some ecosystems can exist in multiple stable states with different species compositions. Shifts between these states can occur due to environmental changes or disturbances.

Empirical Studies

Ecologists conduct field and laboratory studies to test coexistence theories and observe coexisting species in natural ecosystems. These studies provide insights into the factors that promote coexistence.

Conservation Implications

Coexistence theory has implications for conservation biology by highlighting the importance of maintaining diverse habitats, managing disturbance regimes, and considering species interactions in conservation strategies.

Hence, coexistence theory is a fundamental concept in ecology that addresses how multiple species can persist together in the same habitat despite competition for limited resources. It encompasses various mechanisms, from resource partitioning to predator-mediated coexistence, and provides valuable insights into the maintenance of biodiversity and the functioning of ecosystems. Understanding coexistence is essential for effective conservation and management of natural environments.

4.25. COMPETITION, PREDATION, AND PARASITISM

In ecology, biological interactions between species are essential for understanding the dynamics of ecosystems. These interactions can be categorized into various types, each with its own impact on the participating species and the ecosystem as a whole. Here, we explore three specific types of biological interactions: competition, predation, and parasitism.

Competition

Competition is a biological interaction in which two or more species compete for limited resources, such as food, water, space, or nutrients. It can occur within a single species (intraspecific competition) or between different species (interspecific competition).

Examples of Competition

In a forest ecosystem, different tree species may compete for sunlight, nutrients, and water in the same area.

In a pond, multiple species of fish may compete for the same prey, like aquatic insects.

Predation

Predation is a biological interaction in which one species, the predator, captures, kills, and consumes another species, the prey. Predators have evolved various adaptations to capture prey, including sharp claws, teeth, camouflage, and hunting behaviors.

Examples of Predation

A lion hunting and consuming a gazelle on the African savannah.

A spider capturing and feeding on a fly caught in its web.

Parasitism

Parasitism is a biological interaction in which one species, the parasite, benefits at the expense of another species, the host. Parasites can live inside or on their host, extracting nutrients or causing harm to the host organism.

Examples of Parasitism

A tapeworm living in the intestines of a mammal and absorbing nutrients from the host's digestive system.

Fleas infesting a dog and feeding on its blood, causing discomfort and health issues for the dog.

Ecological Significance

Competition: It can influence species distributions, resource use, and population dynamics. Intense competition can lead to competitive exclusion, where one species outcompetes and eliminates another. Resource partitioning and niche differentiation can allow species to coexist.

Predation: Predators help control prey populations and influence prey behavior and evolution. They can also affect community structure and diversity by reducing the abundance of certain species.

Parasitism: Parasites can impact host populations and health. They can also influence the behavior, physiology, and life history of their hosts. Parasites can be important in regulating host populations.

Coexistence: In many ecosystems, these interactions play a crucial role in promoting species coexistence. Predators can reduce competition among prey species, while parasites can influence the competitive abilities of host species. The complex web of interactions in ecosystems helps maintain biodiversity and stability.

In a nutshell, competition, predation, and parasitism are essential biological interactions that shape ecosystems and influence species distributions, population dynamics, and community structure. Understanding these interactions is vital for comprehending the complexity of ecological systems and their conservation and management.

4.26. REFERENCES

1. Abel, N., Cork, S., Gorddard, R., Langridge, J., Langston, A., Plant, R., Proctor, W., Ryan, P., Shelton, D., Walker, B., & Yialeloglou, M., (2003). Natural Values: Exploring Options for Enhancing Ecosystem Services in the Goulburn Broken Catchment, CSIRO, Canberra. Australian State of the Environment Committee (2001). Australia State of the Environment 2001.
2. Beeton, R. J. S., (2006). 'Society's forms of capital: A framework for renewing our thinking.' Paper prepared for the 2006 Australian State of the Environment Committee, Department of the Environment and Heritage, Canberra. www.environment.gov.au/soe/2006/publications/emerging/ capital/index.html
3. Binning, C., Cork, S., Parry, R., & Shelton, D., (2001). Natural Assets: An inventory of Ecosystem Goods and Services in the Goulburn Broken Catchment. CSIRO, Canberra.
4. Cork, S., Stoneham, G., & Lowe, K., (2007). 'Ecosystem services and Australian natural resource management (NRM) futures.' Paper to the Natural Resource Policies and Programs Committee(NRPPC) and the Natural Resource Management Standing Committee (NRMSC).
5. Costanza, R., d'Arge, R., De Groot, R., Farber, S., Grasso, M., Hannon, B., Limburg, K., Naeem, S., O'Neill, R. V., Paruelo, J., Raskin, R. G., Sutton, P., & Vandenbelt, M., (1997). 'The value of the world's ecosystem services and natural capital,' Nature 387, 253–260.
6. Crooks, K. R., & Sanjayan, M., (eds) (2006). Connectivity Conservation, Conservation Biology 14, Cambridge University Press.
7. Department of Prime Minister and Cabinet (2008). Australia 2020 Summit: Final Report.
8. Department of Prime Minister and Cabinet, Canberra.
9. Fischer, J., Lindenmayer, D. B., & Manning, A. D., (2006). 'Biodiversity, ecosystem function, and resilience: ten guiding principles for commodity production landscapes,' Front. Ecol. Environment 4(2),80–86. http://cres.anu.edu.au/dbl/paperjf1.pdf (accessed on 29 September 2023).
10. Folke, C., Carpenter, S., Walker, B., Scheffer, M., Elmqvist, T., Gunderson, L., & Holling, C. S., (2004). 'Regime shifts, resilience, and biodiversity in ecosystem management,' Annual Review of Ecology, Evolution and Systematics 35, 557–581.

11. Gardner, G., & Prugh, T., (2008). 'Seeding the Sustainable Economy' in: 2008 State of the World Report, Worldwatch Institute.
12. Haberl, H., Erb, K-H., & Plutzar, C., (2005). The utility of HANPP as an indicator of socio-economic pressures on Biodiversity. International Human Dimension Programme on Global Environmental Change Open Meeting 2005 Session Biodiversity Conservation. http://www.iff.ac.at/socec.
13. Haines-Young, R., Potschin, M., Groot de, R. S., Keinmast, F., & Bolliger, J., (2009). 'Towards a Common International Classification of Ecosystem Services (CICES) for Integrated Environmental and Economic Accounting.' Report to the European Environment Agency, contract no. EEA/ BSS/7/07.
14. Hughes, T. P., (2008). Human impact on coral reefs in: The Great Barrier Reef: biology, environment and management. CSIRO Publishing, Collingwood, VIC, Australia.
15. Independent Report to the Commonwealth Minister for the Environment and Heritage, CSIRO publishing on behalf of the Department of the Environment and Heritage, Canberra.
16. Mackey, B. G., Keith, H., Berry, L., & Lindenmayer, D. B., (2008). 'Green Carbon, the role of natural forests in carbon storage, Part 1, A green carbon account of Australia's SE Eucalypt forests and policy implications,' ANU E-Press, Canberra.
17. Marshall, P., & Schuttenberg, H., (2006). A Reef Manager's Guide to Coral Bleaching, Great Barrier Reef Marine Park Authority, Townsville, Australia.
18. Michaels, K., Lacey, M., Norton, T., & Williams, J., (2008). Vegetation Futures for Tasmania, Veg Futures Conference, Toowoomba.
19. Millennium Ecosystem Assessment (2005). Ecosystems and Human Well-Being: Our Human Planet: Summary for Decision Makers. The Millennium Ecosystem Assessment Series, Volume 5, Island Press, Washington DC.
20. Molles, M. C. (2016). Ecology: concepts and applications. Seventh edition. New York, NY, McGraw-Hill Education.
21. Mullon, C., Fréon, P., & Cury, P., (2005). 'The dynamics of collapse in world fisheries,' Fish and Fisheries 6(2),111–120.

22. Ridder, B., (2008). 'Questioning the ecosystem services argument for biodiversity conservation,' Biodiversity and Conservation 17(4), 781–790.
23. United States Environmental Protection Agency (USEPA) (2009). 'Valuing the Protection of Ecological Systems and Services.' A Report of the EPA Science Advisory Board, EPA-SAB-09–012, Washington DC.
24. Walker, B., & Salt, D., (2006). Resilience Thinking—Sustaining Ecosystems and People in a Changing World, Island Press, Washington DC.

CHAPTER 5

Classification of Ecology

CONTENTS

5.1. INTRODUCTION

Ecology is a diverse field that encompasses various subdisciplines and classifications based on different criteria. Here's an overview of some well-known typologies and classifications within the field of ecology:

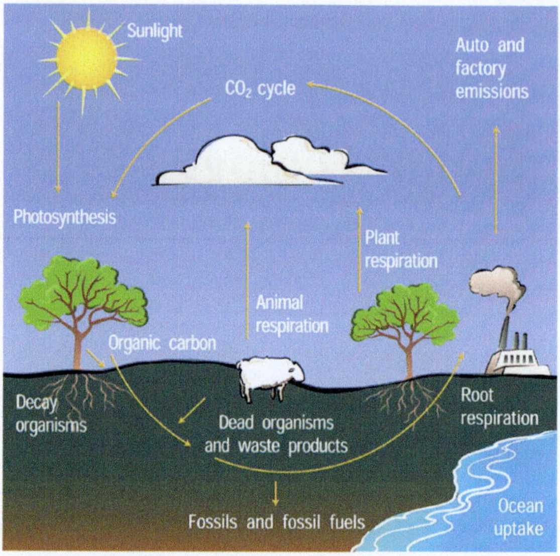

Figure 34. Classification of Ecology

(Source: By Mrs. McComas – http://kmbiology.weebly.com/photosynthesis-cellular-respiration-and-the-carbon-cycle.html, CC BY-SA 4.0, https://commons.wikimedia.org/w/index.php?curid=95286432, Creative Commons License)

Based on Scale or Organizational Levels

- Individual Ecology: Focusing on the study of individual organisms, their behavior, and adaptations.
- Population Ecology: Investigating the dynamics of populations of the same species.
- Community Ecology: Analyzing interactions among different species in a defined area.
- Ecosystem Ecology: Examining the interactions between living organisms and their physical environment.
- Landscape Ecology: Studying the spatial arrangement of ecosystems and their connections in a larger geographic context.

- Global Ecology: Exploring ecological patterns and processes on a global scale, often related to biogeochemical cycles.

Based on Research Focus

- Behavioral Ecology: Investigating the behavior of organisms in response to ecological challenges.
- Conservation Ecology: Concentrating on the preservation and restoration of biodiversity and ecosystems.
- Evolutionary Ecology: Examining how ecological interactions influence evolutionary processes.
- Applied Ecology: Using ecological principles to address real-world environmental problems.
- Theoretical Ecology: Developing mathematical models and theories to explain ecological phenomena.

Based on Habitat or Ecosystem Types

- Aquatic Ecology: Studying ecosystems in water bodies, including oceans, rivers, lakes, and wetlands.
- Terrestrial Ecology: Focusing on ecosystems on land, including forests, grasslands, deserts, and urban areas.
- Marine Ecology: Exploring marine ecosystems and their unique challenges and adaptations.
- Freshwater Ecology: Concentrating on ecosystems in freshwater environments.

Based on Temporal Scales

- Paleoecology: Investigating ancient ecosystems and their changes over geological time.
- Temporal Ecology: Studying ecological processes over various time scales, from seconds to millennia.

Based on Taxonomic Groups

- Avian Ecology: Focusing on the ecology of birds.
- Herpetology: Studying amphibians and reptiles and their ecological roles.

- Insect Ecology: Investigating the ecological interactions of insects.
- Plant Ecology: Concentrating on the ecology of plants and vegetation.

Based on Applied Areas

- Urban Ecology: Examining ecosystems within urban environments.
- Agricultural Ecology: Investigating ecological principles in agriculture and farming systems.
- Industrial Ecology: Analyzing the ecological impact of industrial processes and systems.

Based on Interaction Types

- Predator-Prey Ecology: Studying interactions between predators and their prey.
- Mutualistic Ecology: Investigating mutualistic relationships between species.
- Parasitology: Concentrating on the ecology of parasites and their hosts.
- Competition Ecology: Analyzing competition among species for limited resources.

These classifications provide a framework for organizing and studying ecological concepts and phenomena. Ecologists often specialize in one or more of these typologies to address specific research questions and contribute to our understanding of the natural world.

5.2. ANIMAL ECOLOGY, PLANT ECOLOGY, AND INSECT ECOLOGY

Figure 35. Animal, Plant Ecology, and Insect Ecology

(Source: By Monteuxs – Own work, CC BY-SA 4.0, https://commons.wikimedia.org/w/index.php?curid=54049620, Creative Commons License)

Animal Ecology

Animal ecology is a branch of ecology that focuses on the study of animals and their interactions with their environment.

Key Areas of Study:

- Behavior: Investigating the behavior of animals, including mating, foraging, and migration patterns.
- Population Dynamics: Analyzing the abundance, distribution, and changes in animal populations.
- Community Interactions: Studying how different animal species interact within ecosystems, including competition, predation, and mutualism.
- Conservation: Examining the conservation and management of animal species and their habitats.
- Wildlife Ecology: Concentrating on the ecology of wild animals in natural and human-altered environments.

Examples of Research Topics: Tracking the movement of migratory birds, studying the foraging behavior of wolves, and assessing the impact of human activities on wildlife populations.

Plant Ecology

Plant ecology is a branch of ecology that focuses on the study of plants and their interactions with their environment.

Key Areas of Study:

- Vegetation Dynamics: Analyzing the composition and structure of plant communities and how they change over time.
- Plant Adaptations: Investigating how plants have evolved to survive and thrive in various environmental conditions.
- Ecosystem Function: Examining the role of plants in ecosystem processes, including photosynthesis, nutrient cycling, and carbon sequestration.
- Restoration Ecology: Concentrating on the restoration and management of plant communities in degraded ecosystems.

Examples of Research Topics: Studying the effects of climate change on plant distributions, researching the role of specific plant species in nutrient cycling, and restoring native vegetation in disturbed habitats.

Insect Ecology

Insect ecology is a specialized branch of ecology that focuses on the study of insects (arthropods) and their interactions with their environment.

Key Areas of Study:

- Insect Behaviour: Investigating the behavior of insects, including mating, feeding, and communication.
- Insect-Plant Interactions: Analyzing the relationships between insects and the plants they interact with, including herbivory and pollination.
- Insect Population Dynamics: Studying the abundance and distribution of insect populations and the factors that influence them.
- Pest Management: Concentrating on the ecological management of insect pests in agriculture and forestry.

Examples of Research Topics: Examining the ecological role of pollinators in plant reproduction, studying the impact of invasive insect species on ecosystems, and researching the behavior of social insects like ants and bees.

These subfields within ecology offer a deeper understanding of the ecological interactions and dynamics specific to animals, plants, and insects, contributing to our overall comprehension of ecosystems and biodiversity.

5.3. FOREST ECOLOGY, GRASSLAND ECOLOGY, DESERT ECOLOGY, BENTHIC ECOLOGY, MARINE ECOLOGY, AND URBAN ECOLOGY

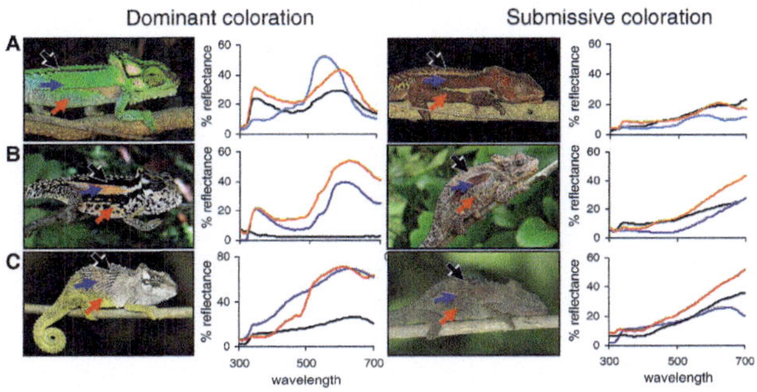

Figure 36. Interesting Studies

(Source: By Stuart-Fox, D., Moussalli, A., – Stuart-Fox, D., Moussalli, A., 2008. Selection for Social Signalling Drives the Evolution of Chameleon Colour Change. PLoS Biol 6(1), e25. doi:10.1371/journal.pbio.0060025, CC BY 2.5, https://commons.wikimedia.org/w/index.php?curid=8870551, Creative Commons License)

Forest Ecology

Forest ecology is the study of forest ecosystems, including the interactions between the biotic and abiotic components of these environments.

Key Areas of Study:

- Forest Structure and Composition: Analyzing the types of trees, understory plants, and wildlife present in different forest types.
- Nutrient Cycling: Investigating how nutrients are cycled through the forest ecosystem, including processes like decomposition.
- Succession: Studying the natural progression of forest ecosystems from early to mature stages.
- Conservation and Management: Concentrating on strategies for sustainable forest management and the conservation of forest biodiversity.

Examples of Research Topics: Assessing the impact of logging on forest ecosystems, studying the role of forests in carbon sequestration, and analyzing the effects of climate change on forest composition.

Grassland Ecology

Grassland ecology is the study of grassland ecosystems, which are characterized by herbaceous vegetation, such as grasses and forbs.

Key Areas of Study:

- Grassland Biodiversity: Analyzing the plant and animal species that inhabit grassland environments.
- Grazing Ecology: Investigating the interactions between herbivores and grassland plants, including the impact of grazing on vegetation.
- Fire Ecology: Studying the role of fire in maintaining grassland ecosystems.
- Restoration Ecology: Concentrating on the restoration of degraded or converted grassland habitats.

Examples of Research Topics: Assessing the effects of overgrazing on grassland plant communities, studying the impact of prescribed burns on grassland health, and restoring native grassland ecosystems.

Desert Ecology

Desert ecology is the study of desert ecosystems, which are characterized by arid conditions, low precipitation, and unique adaptations of organisms to water scarcity.

Key Areas of Study:

- Water Conservation Strategies: Investigating how desert plants and animals have evolved to conserve and utilize water efficiently.
- Extreme Temperatures: Analyzing how organisms cope with extreme temperature fluctuations in desert environments.
- Desert Flora and Fauna: Studying the diverse species that inhabit desert ecosystems, including desert-adapted plants, insects, reptiles, and mammals.
- Desertification: Concentrating on the causes and consequences of desertification and strategies for desert ecosystem conservation.

Examples of Research Topics: Examining the physiological adaptations of desert plants, tracking the movement of desert-dwelling animal species, and assessing the impact of human activities on desert ecosystems.

Benthic Ecology

Benthic ecology is the study of organisms and ecological processes that occur on the seabed or lake bottom, including the study of aquatic sediment ecosystems.

Key Areas of Study:

- Benthic Communities: Analyzing the composition and diversity of organisms living in benthic habitats.
- Sediment Biogeochemistry: Investigating the cycling of nutrients and organic matter in benthic sediments.
- Pollution and Contamination: Studying the impact of pollutants on benthic ecosystems and the health of aquatic organisms.
- Conservation and Restoration: Concentrating on the protection and restoration of benthic habitats.

Examples of Research Topics: Assessing the effects of bottom trawling on benthic communities, studying the role of benthic organisms in nutrient cycling, and developing strategies for the restoration of degraded benthic habitats.

Marine Ecology

Marine ecology is the study of ocean ecosystems, including the interactions between marine organisms and their physical and chemical environments.

Key Areas of Study:

- Coral Reefs: Analyzing the biodiversity and conservation of coral reef ecosystems.
- Pelagic Ecosystems: Investigating open ocean ecosystems and the distribution of plankton, fish, and marine mammals.
- Deep-Sea Ecology: Studying the unique and extreme environments of the deep ocean, including hydrothermal vents and cold seeps.
- Marine Conservation: Concentrating on strategies to protect and preserve marine biodiversity and habitats.
- Examples of Research Topics: Assessing the impact of ocean

acidification on coral reefs, studying the migration patterns of marine species, and exploring the ecology of deep-sea organisms.

Urban Ecology

Urban ecology is the study of ecosystems within urban and metropolitan areas, including the interactions between human activities and the environment.

Key Areas of Study:

- Urban Biodiversity: Analyzing the flora and fauna found in cities and the factors influencing their presence.
- Urban Ecosystem Services: Investigating the benefits provided by urban ecosystems, such as air and water purification, temperature regulation, and recreational spaces.
- Urbanization Impacts: Studying the ecological effects of urban development, including habitat fragmentation, pollution, and altered hydrology.
- Sustainable Urban Planning: Concentrating on strategies for creating more sustainable and resilient urban environments.

Examples of Research Topics: Assessing the impact of green roofs on urban biodiversity, studying urban heat island effects, developing urban wildlife corridors.

These branches of ecology provide insights into the unique dynamics and challenges of different ecosystems, helping us understand and conserve Earth's diverse environments.

5.4. ARCTIC ECOLOGY AND TROPICAL ECOLOGY

Figure 37. Arctic Ecology

(Source: By Harmonia Amanda – Own work, CC BY-SA 3.0, https://commons.wikimedia.org/w/index.php?curid=6786635, Creative Commons License)

Arctic Ecology

Arctic ecology is the study of ecosystems in the Arctic region, which includes the polar and subpolar areas characterized by extremely cold temperatures, ice cover, and unique adaptations of organisms to cold and extreme conditions.

Key Areas of Study:

- Arctic Flora and Fauna: Analyzing the plant and animal species that inhabit the Arctic, including iconic species like polar bears, Arctic foxes, and muskoxen.
- Cold Adaptations: Investigating how organisms have evolved to survive in the extreme cold, including strategies for thermoregulation and food acquisition.
- Sea Ice Ecosystems: Studying the unique ecosystems that develop in and around sea ice, such as algae, phytoplankton, and the organisms that rely on them.
- Climate Change Impacts: Concentrating on the ecological consequences of climate change in the Arctic, including sea ice loss, shifts in species distributions, and altered food webs.
- Examples of Research Topics: Tracking the migration patterns of Arctic seabirds, studying the effects of melting permafrost on vegetation, assessing the vulnerability of Arctic ecosystems to oil spills.

Tropical Ecology

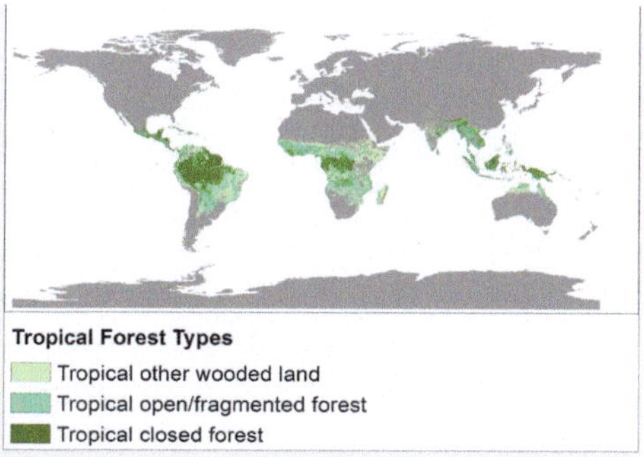

Figure 38. Tropical Ecology

(Source: By Mark Marathon – Own work, CC BY-SA 4.0, https://commons.wikimedia.org/w/index.php?curid=61122968, Creative Commons License)

Tropical ecology is the study of ecosystems in tropical regions, typically located near the equator, characterized by high temperatures, high biodiversity, and unique adaptations of organisms to warm and humid conditions.

Key Areas of Study:

- Rainforest Ecosystems: Analyzing the structure, composition, and biodiversity of tropical rainforests, which are among the most biologically diverse ecosystems on Earth.
- Coral Reef Ecology: Investigating the ecology of coral reefs and the interactions between corals, reef fish, and other marine organisms.
- Tropical Biodiversity: Studying the myriad of species that inhabit tropical ecosystems, including insects, amphibians, reptiles, birds, and mammals.
- Deforestation and Conservation: Concentrating on the impact of deforestation, habitat loss, and conservation efforts in tropical regions.

Examples of Research Topics: Assessing the effects of climate change on tropical bird migration, studying the ecological functions of tropical insects, and evaluating the effectiveness of protected areas in conserving tropical biodiversity.

Arctic and tropical ecosystems represent two extreme ends of the ecological spectrum, each with its own set of challenges and unique adaptations. Researchers in these fields play a crucial role in understanding the ecological dynamics of these environments and in addressing conservation and climate change issues.

5.5. MACROECOLOGY AND LANDSCAPE ECOLOGY

Macroecology

Macroecology is a subfield of ecology that focuses on the study of ecological patterns and processes at large spatial and temporal scales. It involves the analysis of broad-scale ecological phenomena, such as species distributions, biodiversity gradients, and ecosystem dynamics, often using large datasets and statistical models.

Key Areas of Study:

- Species Distributions: Analyzing the global or regional distribution patterns of species and the factors that influence these patterns.
- Biodiversity Gradients: Investigating how species richness and diversity change across geographic regions, climate zones, and habitat types.
- Biogeography: Studying the historical and ecological factors that shape the geographic distribution of species and communities.
- Scaling Laws: Examining the relationships between ecological variables and scale, such as the species-area relationship and the abundance-occupancy relationship.

Examples of Research Topics: Exploring the latitudinal diversity gradient, analyzing the factors driving species richness in ocean ecosystems, and using remote sensing to study large-scale vegetation patterns.

Landscape Ecology

Landscape ecology is a subfield of ecology that focuses on the study of spatial patterns and processes in landscapes, considering the interactions between ecological systems, land use, and human activities. It examines how landscape structure influences ecological functions and biodiversity.

Key Areas of Study:

- Landscape Structure: Analyzing the arrangement and composition of different landscape elements, such as forests, wetlands, urban areas, and agricultural fields.
- Habitat Fragmentation: Investigating the effects of habitat loss and fragmentation on wildlife populations and ecological connectivity.
- Ecological Connectivity: Studying the flow of organisms, materials, and energy across landscapes, including corridors and barriers.
- Landscape Change and Conservation: Concentrating on land-use changes, such as urbanization, deforestation, and agricultural expansion, and their implications for biodiversity and ecosystem services.
- Examples of Research Topics: Assessing the impact of urban

sprawl on local wildlife habitats, studying the design of wildlife corridors to promote genetic diversity, modelling the effects of land use on water quality in a watershed.

Both Macroecology and Landscape Ecology deal with the analysis of ecological phenomena at larger scales and emphasize the importance of understanding spatial and temporal patterns in ecological systems. Macroecology takes a broad, global perspective on ecological patterns, while Landscape Ecology focuses on the structure and function of landscapes and their influence on local ecosystems. Both subfields contribute to our understanding of how ecosystems function on a large scale and are essential for informing conservation and land management practices.

5.6. SYSTEMS ECOLOGY

Systems ecology is a branch of ecology that views ecological systems as complex, interconnected networks of organisms and their environment, where the focus is on understanding the relationships, feedback loops, and emergent properties of these systems. It employs a holistic approach to studying the dynamics and behavior of ecological systems.

Key Areas of Study:

- System Dynamics: Analyzing the dynamic behavior of ecological systems, including the flow of energy, nutrients, and information within and between components of the system.
- Ecosystem Modelling: Developing mathematical models and simulations to represent and predict the behavior of ecological systems, allowing for scenario testing and hypothesis exploration.
- Feedback Mechanisms: Investigating the various feedback loops within ecological systems, including positive feedback that amplifies changes and negative feedback that stabilizes them.
- Emergent Properties: Studying the properties of ecological systems that arise from the interactions of their components, often leading to unexpected behaviors or patterns.
- Network Analysis: Using tools from network theory to understand the structure of ecological networks, including food webs, species interactions, and mutualistic relationships.
- Resilience and Stability: Assessing the resilience of ecological systems to disturbances and the factors that contribute to system stability or vulnerability.

Examples of Research Topics:

- Modeling the dynamics of predator-prey relationships in ecosystems.
- Analyzing the impact of climate change on the resilience of coral reef ecosystems.
- Studying the feedback loops in forest ecosystems, including how deforestation can alter precipitation patterns.
- Developing models to predict the spread of invasive species within aquatic ecosystems.
- Investigating the flow of energy and nutrients in freshwater food webs.

Systems ecology recognizes the complexity and interconnectedness of ecological systems and seeks to provide a deeper understanding of how these systems function, respond to change, and can be managed for sustainability and conservation. It often involves interdisciplinary collaboration with mathematics, computer science, and other fields to develop and apply models and analytical tools for studying ecological systems.

5.7. MOLECULAR ECOLOGY

Molecular ecology is a specialized subfield of ecology that uses genetic and molecular techniques to investigate the ecological interactions, behaviors, and population dynamics of organisms. It focuses on understanding how genetic information can provide insights into the ecology, evolution, and conservation of species and ecosystems.

Key Areas of Study:

- Population Genetics: Analyzing genetic variation within and among populations to study factors like gene flow, genetic diversity, and the genetic structure of populations.
- Phylogeography: Examining the genetic and evolutionary history of populations and species to understand their historical distribution patterns and diversification.
- Genomic Ecology: Utilizing advanced genomic technologies, such as DNA sequencing, to study the genetic basis of ecological adaptations, phenotypic traits, and responses to environmental changes.
- Conservation Genetics: Applying genetic data to inform

conservation efforts, including the identification of endangered populations, management of genetic diversity, and strategies for captive breeding and reintroduction programs.

- Behavioural Ecology: Using molecular tools to investigate the genetic basis of behaviours, mating systems, and social interactions within populations.

Examples of Research Topics:

- Studying the genetic diversity and population structure of endangered species to inform conservation strategies.
- Investigating how climate change affects the genetic adaptation and distribution of species.
- Using DNA barcoding to identify species and track their movements in ecological communities.
- Analyzing the genetic basis of migratory behaviors in birds and other animals.
- Examining the impact of human activities on the genetic diversity of wildlife populations.

Molecular ecology combines principles of genetics, genomics, and ecology to provide a deeper understanding of the genetic basis of ecological processes and the evolutionary forces that shape ecosystems. It plays a crucial role in wildlife conservation, the study of adaptation to changing environments, and the preservation of biodiversity.

5.8. REFERENCES

1. Abramovitz, J. N. (1991). Investing in Biological Diversity: U.S. Research and Conservation Efforts in Developing Countries. Washington, D.C.: World Resources Institute.
2. Alcorn, J., & C. Hernàndez, V. (1983). Plants of the Huastecan region of Mexico with analysis of their Huastec names. J. Mayan Ling. 4, 11–118.
3. Allen, T. F. H., & Starr, T. B. (1982). Hierarchy: Perspectives for Ecological Complexity. Chicago: University of Chicago Press.
4. Amaranthus, M. P., & Perry, D. A. (1987). The effect of soil inoculation on ectomycorrhizal formation and the survival and growth of conifer seedlings on old, non-reforested clearcuts. Can. J. For. Res. 17, 944–950.
5. Ashton, P. (1989). Funding Priorities for Research Towards Effective Sustainable Management of Biodiversity Resources in Tropical Asia. Report of a workshop sponsored by NSF and USAID held in Bangkok, Thailand, March 27–30, (1989). Unpublished report.
6. Berlin, B., Breedlove, D. E., & Raven, P. (1974). Principles of Tzeltal Plant Classification: An Introduction to the Botanical Ethnography of a Mayan-Speaking People of Highland Chiapas. New York: Academic Press.
7. Böhlke, J. E., Weizman, S. H., & Manezes, N. A. (1978). Estado atual da sistemática dos peixes de água doce da América do Sul. Acta Amazonica 8(4),657–677.
8. Bormann, F. H., & Likens, E. G. (1979). Pattern and Process in a Forested Ecosystem. New York: Springer-Verlag.
9. Braden, J. B., & Kolstad, D. C. (eds.) (1991). Measuring the Demand for Environmental Quality. New York: Elsevier.
10. Brush, S. B., Carney, H. J., & Muamán, Z. (1991). Dynamics of Andean potato agriculture. Economic Botany 35, 70–88.
11. Burley, W. F. (1988). The tropical forest action plan: Recent progress and new initiatives. pp. 403–408 in Biodiversity, E. O. Wilson, & Peter, M. F. (eds.) Washington, D.C.: National Academy Press.
12. Cairns, Jr. J., (1988). Increasing diversity by restoring damaged ecosystems. pp. 333–343 in Biodiversity, E. O. Wilson, & Peter, M. F. (eds.) Washington, D.C.: National Academy Press.

13. Charney, J. G. (1975). Dynamics of deserts and drought in the Sahel. Quarterly Journal of the Royal Meteorological Society 101 (428):193–202.
14. de Chassey, F. (1978). La Mauritanie—1900–1975. Paris: Editions Anthropos.
15. Christonty, L. (1986). Shifting cultivation and tropical soils: Patterns, problems, and possible improvements. pp. 226–240 in Traditional Agriculture in South East Asia: A Human Ecology Perspective, G. G. Martin, ed. Honolulu, Hawaii: East-West Center.
16. Conklin, H. C. (1980). Ethnographic Atlas of Ifugao. New Haven, Conn.: Yale University Press.
17. Cook, R. J. (1991). Challenges and rewards of sustainable agriculture research and education. pp. 32–76 in Sustainable Agriculture Research and Education in the Field. Washington, D.C.: National Academy Press.
18. Costanza, R., ed. (1991). The Science and Management of Sustainability. New York: Columbia University Press.
19. Costanza, R., & Perrings, C. (1990). A flexible assurance bonding system for improved environmental management. Ecological Economics 2, 57–76.
20. Cremoux, P. (1963). The importance of game-meat consumption in the diet of sedentary and nomadic peoples of the Senegal River Valley. pp. 127–129 in Conservation of Nature and Natural Resources in Modern African States. IUCN Publications New Series, No. 1. Morges, Switzerland: IUCN.
21. Daly, H. E., & Cobb, Jr. B. J., (1990). For the Common Good: Redirecting the Economy Toward Community, the Environment, and a Sustainable Future. Boston: Beacon Press.
22. Demain, A. L., & Solomon, N. A. (1981). Industrial microbiology. Scientific American 245(3),66–75.
23. Ecological Society of America (ESA). (1991). The sustainable biosphere initiative: An ecological research agenda. Ecology 72(2), 371–412.
24. Ehrlich, P. R., & Wilson, E. O. (1991). Biodiversity studies: Science and policy. Science 253, 758–762.
25. Erwin, T. L. (1982). Tropical forests: Their richness in Coleoptera and other arthropod species. Coleopterists' Bulletin 36, 74–75.

26. Erwin, T. L. (1983). Tropical forest canopies: The last biotic frontier. Bulletin of the Entomological Society of America 29(1), 14–19.
27. Erwin, T. L. (1991). How many species are there? Revisited. Conservation Biology 5, 330–333.
28. Franklin, J. F. (1989). Importance and justification of long-term studies in ecology. PP. 3–19 in Long-Term Studies in Ecology: Approaches and Alternatives, G. E. Likens, ed. New York: Springer-Verlag.
29. Franklin, J. F., Bledsoe, C. S., & Callahan, J. T. (1990). Contributions of long-term ecological research program: An expanded network of scientists, sites, and programs can provide crucial comparative analyses. Bioscience 40(7), 509–523.
30. Gaston, A., & Dulieu, D. (1976). Pâturages du Kanem. Maisons-Alfort, France: Institut d'Élevage et de Médecine Vétérinaire des Pays Tropicaux.
31. Gaston, K. J. (1991). The magnitude of global insect species richness. Conservation Biology 5(3),283–296.
32. Gómez-Pompa, A. (1987). On Maya silviculture. Mexican Studies/ Estudios Mexicanos 3(1),1–17.
33. Gómez-Pompa, A. (1988). Tropical deforestation and Maya silviculture: An ecological paradox. Tulane Studies in Zoology and Botany 26, 19–37.
34. Grainger, A. (1984). Quantifying changes in forest cover in the humid tropics: Overcoming current limitations. Journal of World Forest Resource Management 1, 3–63.
35. Grainger, A. (1988). Estimating areas of degraded tropical lands requiring replenishment of forest cover. International Tree Crops Journal 5, 31–61.
36. Green, K. M. (1981). Digital processing of tropical forest habitat in Bangladesh and the development of low cost processing facility at the National Zoo, Smithsonian Institution. Proceedings of 15th International Symposium on Remote Sensing of Environment 3, 1315–1325.
37. Green, K. M., Lynch, J. F., Sircar, J., & Greenberg, Z. L. S. (1987). Landsat remote sensing to assess habitat for migratory birds in the Yucatan peninsula, Mexico. Vida Silvestre Neotropical 1(2),27–38.
38. Guzman, H. M. (1991). Restoration of coral reefs in Pacific Costa Rica. Conservation Biology 5(2),189–195.

39. Harley, J. L., & Smith, S. E. (1983). Mycorrhizal Symbiosis. New York: Academic Press.
40. Harris, L. D. (1984). The Fragmented Forest: Island Biogeography Theory and the Preservation of Biotic Diversity. Chicago: University of Chicago Press.
41. Heady, H. F., & Heady, E. B. (1982). Range and Wildlife Management in the Tropics. London: Longman Press.
42. Houghton, J. T., Jenkins, G. J., & J, J. Ephraums, (eds.) (1990). Climate Change: The IPCC Scientific Assessment. Cambridge, England: Cambridge University Press.
43. Jablonski, D. (1991). Extinctions: A paleontological perspective. Science 253, 754–757.
44. Janzen, D. H. (1988). Tropical dry forests: The most endangered major tropical ecosystem. pp. 130–137 in Biodiversity, E. O. Wilson, & Peter, M. F. (eds.) Washington, D.C.: National Academy Press.
45. Jenkins, Jr., R. E. (1988). Heritage Conservation Data Center. Paper presented to IUCN General Assembly, February 4, 1988.
46. Jordan III, W. R. (1988). Ecological restoration: Reflections of a half-century of experience at the University of Wisconsin-Madison Arboretum. pp. 311–316 in Biodiversity, E. O. Wilson, & Peter, M. F. (eds.) Washington. D.C.: National Academy Press.
47. Jordan III, W. R., Gilpin, M. E., & Aber, J. D. (1987). Restoration Ecology: A Synthetic Approach to Ecological Research. New York: Cambridge University Press.
48. Kneese, A. V. (1984). Measuring the Benefits of Clean Air and Water. Washington, D.C.: Resources for the Future.
49. Kneese, A. V., & Bower, B. T. (1968). Managing Water Quality: Economics, Technology, Institutions. Washington, D.C.: Resources for the Future.
50. Kristensen, R. M. 1983, Loricifera, a new phylum with Aschelminthes characters from the meiobenthos. Z. Zool. Syst. 21(3), 163–180.
51. Likens, G. E., ed. (1989). Long-Term Studies in Ecology: Approaches and Alternatives. New York: Springer-Verlag.
52. Lorini, M. L., & Persson, V. G. (1990). New species of Leontopithecus lesson 1840 from Southern Brazil primates Callitrichidae. Bol. Mus. Nac. Rio. J. Zool. O. 338, 1–14.

53. Luoma, J. R. (1991). A wealth of forest species is underfoot. New York Times, July 2, (1991). C1; C9.
54. Malhotra, K. C., Khomne, S. B., & Gadgil, M. (1983). Hunting strategies among three non-pastoral nomadic groups of Maharashtra. Man in India 63(1), 23–39.
55. Martin, P. S. (1984). Prehistoric overkill: The global model. pp. 354–403 in Quaternary Extinctions: A Prehistoric Revolution, P. S. Martin, & G, R. Klein (eds.) Tucson, Ariz.; University of Arizona Press.
56. Mayr, E. (1982). The Growth of Biological Thought: Diversity, Evolution, & Inheritance. Cambridge, Mass.: Belknap Press.
57. McNeely, J. A. (1988). Economics and Biological Diversity: Developing and Using Economic Incentives to Conserve Biological Resources. Gland, Switzerland: IUCN.
58. McNeely, J. A. (1989). Priorities in conservation biology. Conservation Biology 3, 416–417.
59. McNeely, J. A., Miller, K. R., W. V. Reid, R. A. Mittermeier, & Werner, T. B. (1990). Conserving the World's Biological Diversity. Gland, Switzerland and Washington, D.C.: IUCN, WRI, CI, WWF-US, and the World Bank.
60. Matson, M., & Holben, B. (1987). Satellite detection of tropical burning in Brazil. International Journal of Remote Sensing 8(3),509–516.
61. May, R. M. (1988). How many species are there on the Earth? Science 241, 1441–1449.
62. Miller, R. M. (1985). Mycorrhizae. Restoration and Management Notes 3(1),14–20.
63. Mitchell, R. C., & Carson, R. T. (1989). Using Surveys to Value Public Goods: The Contingent Valuation Method. Washington, D.C.: Resources for the Future .
64. Morin, N. R., Whetsonte, R. D., Wilken, D., & Tomlinson, K. L. (1989). Floristics for the 21st Century: Proceedings of the Workshop 4–7 May, 1988, Alexandria, Va. Monographs in Systematic Botany from the Missouri Botanical Garden. Volume 28. Ann Arbor, Mich.: Braun-Brumfield.
65. Myers, N. (1988). Threatened biotas: Hotspots in tropical forests. Environmentalist 8(3),1–20.
66. National Academy of Sciences (NAS). (1980). Research Priorities in Tropical Biology. Washington, D.C.: National Academy of Sciences.

67. National Research Council (NRC). (1975). Underexploited Tropical Plants with Promising Economic Value. BOSTID Report 16. Washington, D.C.: National Research Council.
68. National Research Council. (1979). Tropical Legumes: Resources for the Future. BOSTID Report 25. Washington, D.C.: National Research Council.
69. National Research Council. (1982). Ecological Aspects of Development in the Humid Tropics. Washington, D.C.: National Academy Press.
70. National Research Council. (1983). Environmental Change in the West African Sahel. Washington, D.C.: National Academy Press.
71. National Research Council. (1986a). Ecological Knowledge and Environmental Problem-Solving: Concepts and Case Studies. Washington, D.C.: National Academy Press.
72. National Research Council. (1986b). Proceedings of the Conference on Common Property Resource Management. Washington, D.C.: National Academy Press.
73. National Research Council. (1989). Science and Technology Information Services and Systems in Africa. Washington, D.C.: National Academy Press.
74. National Research Council. (1990). The Improvement of Tropical and Subtropical Rangelands. Washington, D.C.: National Academy Press.
75. National Research Council. (1991a). Managing Global Genetic Resources: Forest Trees. Washington, D.C.: National Academy Press.
76. National Research Council. (1991b). Managing Global Genetic Resources: The, U.S. National Plant Germplasm System. Washington, D.C.: National Academy Press.
77. National Research Council. (1991c). Policy Implications of Greenhouse Warming. National Research Council. Washington, D.C.: National Academy Press.
78. National Research Council. (1991d). Towards Sustainability: A Plan for Collaborative Research on Agriculture and Natural Resource Management. Washington, D.C.: National Academy Press.
79. National Science Board (NSB). (1989). Loss of Biological Diversity: A Global Crisis Requiring International Solutions. Report NSB-89–171. Washington: D.C.: National Science Foundation.
80. Nepstad, D.C., Uhl, C., & Serrao, A. S. (1991). Recuperation of a degraded Amazonian landscape: Forest management and agricultural

restoration. Ambio 20(6), 248–255.

81. Neuenschwander, P., Hammond, W. N. O., Ajuonu, O., Gado, A., Eschendu, N., A. H. Bokonon-Ganta, R. Allomasso, & Okon, I. (1990). Biological control of the cassava mealybug, Phenacoccus manihoti (Hom., Pseudococcidae) by Epidinocarsis lopezi (Hym., Encyrtidae) in West Africa, as influenced by climate and soil. Agriculture, Ecosystems, & Environment 32, 39–55.
82. Norgaard, R. (1988). The biological control of cassava mealybug in Africa. American Journal of Agricultural Economics 70(2), 366–371.
83. Norgaard, R. (1991). Sustainability as Intergenerational Equity: The Challenge to Economic Thought and Practice. World Bank Discussion Paper. Washington, D.C.: World Bank.
84. Norse, E. A. (1990). Ancient Forests of the Pacific Northwest. Washington, D.C.: Island Press.
85. Noss, R. F. (1990). Indicators for monitoring biodiversity: A hierarchical approach. Conservation Biology 4(4),355–364.
86. O'Neill, R. V., DeAngelis, D. L., Waide, J. B., & H, T. F. Allen. (1986). A Hierarchical Concept of Ecosystems. Princeton, N. J.: Princeton University Press.
87. Orians, G. H., ed. (1991). The Preservation and Valuation of Biological Resources. Seattle, Wash.: University of Washington Press.
88. Padoch, C. (1986). Site selection among permanent-field farmers: An example in East Kalimantan,
89. Panayotou, T. (1989). Economics of Environmental Degradation Problems, Causes and Responses. Harvard Institute for International Development Report prepared for the, U.S. Agency for International Development under CAER Task Order #3.
90. Panayotou, T. (1990). Counting the cost: Resource degradation in the developing world. Fletcher Forum of World Affairs 14(2), 270–283.
91. Pearce, D. W., & Turner, R. K. (1990). Economics of Natural Resources and the Environment. Baltimore, Md.: Johns Hopkins Press.
92. Peters, C. M., Gentry, A. H., & Mendelsohn, R., (1989). Valuation of a tropical forest in Peruvian Amazonia. Nature 339, 655–656.
93. Population Reference Bureau (PRB). (1989). World Population Data Sheet (computer diskette). Washington, D.C.: Population Reference Bureau.

94. Pyne, S. J. (1991). Burning Bush: A Fire History of Australia. New York: Henry Holt.
95. Randall, A. (1991a). Nonuse Benefits. In: Measuring the Demand for Environmental Quality, J. B. Braden, & Kolstad, D. C. (eds.) New York: Elsevier.
96. Randall, A. (1991b). Thinking about the value of biodiversity. Unpublished manuscript.
97. Raven, P. (1988). Biological resources and global stability. pp. 3–27 in Evolution and Coadaptation in Biotic Communities, Kawano, S., J. H. Connell, & Hidaka, T., (eds.) Tokyo: University of Tokyo Press.
98. Ray, G. C. (1988). Ecological diversity in coastal zones and oceans. pp. 36–50 in Biodiversity, E. O. Wilson, & Peter, M. F. (eds.) Washington, D.C.: National Academy Press.
99. Red Latinoamericana de Botànica (RLB). (1991). Report of Activities, (1990). Santiago: RLB.
100. Reid, W. V., & Miller, K. R. (1989). Keeping Options Alive: The Scientific Base for Conserving Biodiversity. Washington, D.C.: World Resources Institute.
101. Repetto, R. (1988). The Forest for the Trees? Government Policies and the Misuse of Forest Resources. Washington, D.C.: World Resources Institute.
102. Repetto, R., & Gillis, M., (eds.) (1988). Public Policies and the Misuse of Forest Resources. Cambridge, England: Cambridge University Press.
103. Risser, P. G., & Melillo, J. M. (1991). LTER: An International Proposal. New York: John Wiley & Sons.
104. Roberts, L. (1991). Ranking the rain forest. Science 251, 1559–1560.
105. Robinson, M. H. (1988). Are there alternatives to destruction? pp. 355–360 in Biodiversity, E. O. Wilson, & Peter, M. F. (eds.) Washington, D.C.: National Academy Press.
106. Sader, S. A., Stone, T. A., & Joyce, A. T. (1990). Remote sensing of tropical forests: Overview of research and applications using non-photographic sensors. Photogrammetric Engineering and Remote Sensing 56(10):1343–1351.
107. Salam, A. (1989). Notes on Science, Technology and Science Education in the Development of the South. Trieste, Italy: Third World Academy of Sciences.

108. Sale, J. B. (1981). The Importance and Values of Wild Plants and Animals in Africa. Gland, Switzerland: IUCN.

109. Salvat, B. (1987). Human Impacts on Coral Reefs: Facts and Recommendations. French Polynesia: Antennes Museum, E. P. H. E.

110. Saunders, D. A., Hobbs, R. J., & Margules, C. R. (1991). Biological consequences of ecosystem fragmentation: A review. Conservation Biology 5(1),18–32.

111. Scott, J. M., Csuti, B., & Davis, F. (1991a). Gap analysis: An application of global information systems for wildlife species. pp. 167–180 in Challenge in the Conservation of Bioresources: A Practitioner's Guide, O. J. Decker, M. E. Krasny, F. R. Groff, C. R. Smith, & Gross, W. D. (eds.) Boulder, Colo.: Westview.

112. Scott, J. M., Csuti, B., Smith, K., J. E. Estes, S. Caicco. (1991b). Gap analysis of species richness and vegetation cover: An integrated biodiversity strategy. pp. 282–297 in Balancing on the Brink of Extinction, Kathryn, A. Kohm, ed. Washington, D.C.: Island Press.

113. Solbrig, O., ed. (1991). From Genes to Ecosystems: A Research Agenda for Biodiversity. Cambridge, Mass.: IUBS, SCOPE, & UNESCO.

114. Sondaar, P. Y. (1977). Insularity and its effect on mammal evolution. pp. 671–707 in Major Patterns in Vertebrate Evolution, M. K. Hecht, R. C. Goody, & Hecht, M. B. (eds.) New York: Plenum.

115. Soulé, M. E. (1986). Conservation Biology: The Science of Scarcity and Diversity. Sunderland, Mass.: Sinauer Associates.

116. Soulé, M. E. (1991). Conservation: Tactics for a constant crisis. Science 253, 744–750.

117. Soulé, M. E., & Kohm, A. K. (eds.) (1989). Research Priorities for Conservation Biology. Washington, D.C.: Island Press.

118. Soulé, M. E., & Wilcox, A. B. (eds.) (1980). Conservation Biology: An evolutionary-ecological perspective. Sunderland, Mass.: Sinauer Associates.

119. Stork, N. E. (1988). Insect Diversity: Facts, fiction, and speculation. Biological Journal of the Linnean Society. 35, 321–337.

120. Tangley, L. (1990). Cataloguing Costa Rica's diversity. Bioscience 40, 633–636.

121. Tatum, L. A. (1971). The southern corn leaf blight epidemic. Science 171, 1113–1116.

122. Thomas, G. W. (1980). The Sahelian/Sudanian Zones of Africa: Profile of a Fragile Environment. Report to the Rockefeller Foundation. New York: Rockefeller Foundation.
123. Thorne-Miller, B., & Cantena, J. (1991). The Living Ocean: Understanding and Protecting Marine Biodiversity. Washington, D.C.: Island Press.
124. Uhl, C. (1988). Restoration of degraded lands in the Amazon basin. pp. 326–332 in Biodiversity, E. O. Wilson, & Peter, M. F. (eds.) Washington, D.C.: National Academy Press.
125. Uhl, C., Nepstad, D., Buschbach, R., Clark, K., Kauffman, B., & Subler, S. (1990). Studies of ecosystem response to natural and anthropogenic disturbances provide guidelines for designing sustainable land-use systems in Amazonia. pp. 24–42 in Alternatives to Deforestation: Steps toward Sustainable Use of the Amazon Rain Forest, A. B. Anderson, ed. New York: Columbia University Press.
126. United Nations Educational Scientific, & Cultural Organization (Unesco). (1974). Task Force on Criteria and Guidelines for the Choice and Establishment of Biosphere Reserves. Final Report. MAB Report Series No. 22. Paris: Unesco.
127. U.S. Congress, Office of Technology Assessment (OTA). (1987). Technologies to Maintain Biological Diversity. Washington, D.C.: U.S. Government Printing Office.
128. U.S. Environmental Protection Agency (EPA). (1985). A Methodological Approach to an Economic Analysis of the Beneficial Outcomes of Water Quality. Washington, D.C.: U.S. Environmental Protection Agency.
129. Vibulsresth, S. (1986). Remote sensing activities in Thailand. pp. 127–138 in Remote-Sensing Yearbook. London: Taylor & Francis.
130. Water Resources Council. (1980). Principles and Guidelines for Planning and Related Land Resources. Washington, D.C.: U.S. Government Printing Office.
131. Wilson, E. O. (1987). The arboreal ant fauna of Peruvian Amazon forests: A first assessment. Biotropica 2, 245–251.
132. Wilson, E. O. (1988). The current state of biological diversity. pp. 3–18 in Biodiversity, E. O. Wilson, & Peter, M. F. (eds.) Washington, D.C.: National Academy Press.

133. Wilson, E. O. (1989). Threats to biodiversity. Scientific American 261(3), 108.

134. Wilson, E. O., & Peter, M. F. (eds.) (1988). Biodiversity. Washington, D.C.: National Academy Press.

135. World Bank. (1989). Sub-Saharan Africa: From Crisis to Sustainable Growth. Washington, D.C.: World Bank.

136. World Commission on Environment and Development (WCED). (1987). Our Common Future. Oxford, England: Oxford University Press.

137. World Resources Institute (WRI). (1986). World Resources (1986). New York: Basic Books.

138. World Resources Institute. (1987). World Resources (1987). New York: Basic Books.

139. World Resources Institute. (1991). Compact for a New World. Washington, D.C.: World Resources Institute.

140. National Academies of Sciences, Engineering, & Medicine. (1992). Conserving Biodiversity: A Research Agenda for Development Agencies. Washington, DC: The National Academies Press. Creative Commons License).

CHAPTER 6

Major Subdisciplines of Ecology and Their Further Classifications

CONTENTS

6.1. INTRODUCTION

Ecology is a diverse field that encompasses numerous subdisciplines, each focusing on specific aspects of the relationships between organisms and their environment. Here's an overview of some major subdisciplines of ecology and their further classifications:

Terrestrial Ecology

- Plant Ecology: The study of plant communities, including their distribution, composition, and interactions.
- Animal Ecology: The study of animal species, their behaviors, population dynamics, and interactions with other organisms and their environment.
- Microbial Ecology: Focusing on the ecology of microorganisms, including bacteria, fungi, and archaea, in terrestrial environments.

Aquatic Ecology:

- Limnology: The study of freshwater ecosystems, including lakes, rivers, and wetlands.
- Marine Ecology: The study of marine ecosystems, including oceans, seas, and coastal environments.
- Estuarine Ecology: Focusing on the unique ecosystems found in the transitional zones where freshwater meets saltwater.

Population Ecology:

- Community Ecology: Examining the interactions among different species within a defined geographic area or habitat.
- Ecosystem Ecology: Investigating the flow of energy and nutrients within ecosystems, including the abiotic and biotic components.
- Landscape Ecology: Analyzing the spatial arrangement of ecosystems and their connections across larger geographic areas.

Evolutionary Ecology:

- Behavioral Ecology: Studying the behaviors of organisms in relation to their ecological roles, including mating, foraging, and social interactions.
- Phylogenetic Ecology: Focusing on the evolutionary history of species and how it influences their ecological traits and interactions.
- Functional Ecology: Investigating the functional roles of species

and their adaptations to ecological niches.

Conservation Ecology:

- Restoration Ecology: Focusing on the rehabilitation of ecosystems that have been degraded or damaged.
- Conservation Biology: The study of preserving and protecting biodiversity and threatened species.
- Applied Ecology: Using ecological principles to address practical environmental issues, such as habitat management and sustainable resource use.

Chemical Ecology:

- Plant-Animal Interactions: Studying chemical signals and interactions between plants and herbivores, pollinators, and other organisms.
- Predator-Prey Interactions: Investigating the chemical cues and defenses used by predators and prey in ecological interactions.

Social Ecology:

- Human Ecology: The study of the interactions between humans and their environment, including the impact of human activities on ecosystems.
- Urban Ecology: Focusing on the ecological dynamics within urban environments and the study of cities as ecosystems.

Molecular Ecology:

- Population Genetics: Analyzing genetic variation within populations and the factors influencing gene flow and genetic diversity.
- Phylogeography: Studying the historical distribution and genetic structure of populations and species.

Systems Ecology:

- Ecosystem Modelling: Developing mathematical models to simulate ecological processes and predict ecosystem behavior.
- Network Ecology: Studying the ecological networks and interactions among species within ecosystems.

Each of these subdisciplines may have further subdivisions, and the boundaries between them are often fluid, with interdisciplinary research playing an increasingly significant role in advancing ecological knowledge. Ecologists within these subdisciplines collaborate to provide a comprehensive understanding of the natural world and address critical ecological challenges.

6.2. BY METHODOLOGY USED FOR INVESTIGATION

Field Ecology

A. Population Sampling: Studying organisms in their natural habitats through field observations, surveys, and data collection.

B. Experimental Field Studies: Conducting controlled experiments in the field to manipulate and observe ecological variables.

C. Long-term Ecological Research (LTER): Investigating ecological phenomena over extended time periods to understand long-term trends and patterns.

D. Fieldwork in Extreme Environments: Examining ecosystems in challenging environments like deserts, polar regions, or deep-sea habitats.

Quantitative Ecology

A. Statistical Ecology: Applying statistical methods and modeling techniques to analyze ecological data, including species distributions, population dynamics, and community structure.

B. Mathematical Modeling: Developing and using mathematical models to simulate ecological processes, predict ecological patterns, and test hypotheses.

C. Geospatial Ecology: Utilizing Geographic Information Systems (GIS) and remote sensing technologies for ecological research, such as habitat mapping and spatial analysis.

D. Ecological Data Analysis: Employing advanced data analysis methods to derive ecological insights from complex datasets.

Theoretical Ecology

A. Ecological Modeling: Developing theoretical models to explore ecological phenomena and generate testable hypotheses.

B. Population Dynamics Modeling: Investigating the dynamics of populations using mathematical models, including exponential and logistic growth models.

C. Ecosystem Simulation: Creating computer simulations to study the interactions and feedbacks within ecosystems.

D. Game Theory in Ecology: Applying game theory principles to study ecological interactions, such as predator-prey dynamics and cooperation among species.

These classifications based on investigation methodologies demonstrate the diversity of approaches ecologists use to study and understand the complex ecological processes and relationships in natural systems. Ecologists often combine these methods to gain a comprehensive understanding of ecosystems and their inhabitants.

6.3. BY SPATIAL SCALE OF ECOLOGICAL SYSTEM UNDER STUDY

Global Ecology

Global ecology is the subdiscipline of ecology that focuses on understanding ecological patterns and processes at the largest spatial scale, encompassing the entire planet. It seeks to investigate global ecological phenomena, including climate change, biogeochemical cycles, and the distribution of species on a global scale.

Key Areas of Study:

A. Global Climate Change: Analyzing the impact of human activities on Earth's climate systems and the consequences for ecosystems.

B. Global Biogeochemical Cycles: Investigating the movement of elements like carbon, nitrogen, and phosphorus through Earth's ecosystems.

C. Biodiversity Patterns: Studying the distribution of species and ecosystems worldwide, including hotspots of biodiversity and global conservation efforts.

D. Examples of Research Topics: Assessing the global carbon budget, tracking the movement of migratory species across continents, and studying the effects of ocean currents on global marine ecosystems.

Macroecology

Macroecology is a subfield of ecology that focuses on ecological patterns and processes at large spatial scales, typically involving regional, continental, or biogeographic perspectives. It aims to understand how ecological phenomena vary across broad geographic areas and how environmental factors influence species distributions and diversity.

Key Areas of Study:

A. Biogeography: Analyzing the distribution of species, communities, and ecosystems at regional and continental scales.

B. Species Diversity Gradients: Investigating how species richness changes with latitude, elevation, and other environmental factors.

C. Large-Scale Ecological Patterns: Studying phenomena like island biogeography, species-area relationships, and the latitudinal diversity gradient.

D. Examples of Research Topics: Exploring the factors that drive the high biodiversity of tropical rainforests, examining the biogeography of island species, and studying continental-scale migrations of animals.

Microecology

Microecology is a subfield of ecology that focuses on ecological patterns and processes at very small spatial scales, often involving the study of individual organisms or microhabitats within ecosystems. It investigates the interactions between microorganisms and their environments and how these interactions influence ecosystem dynamics.

Key Areas of Study:

A. Microbial Ecology: Analyzing the diversity and functions of microorganisms (bacteria, fungi, protists) in various ecosystems.

B. Microhabitat Ecology: Investigating ecological interactions within specific microenvironments, such as soil, leaf surfaces, or microbial mats.

C. Microbial Biogeochemistry: Studying the role of microorganisms in nutrient cycling, decomposition, and other biogeochemical processes.

D. Examples of Research Topics: Examining the microbial communities in soil and their contributions to plant health, studying the microbial diversity in extreme environments like hot springs, and investigating the role of microorganisms in carbon sequestration.

These classifications based on spatial scale reflect the varying scopes of ecological research, from the global scale of Earth's ecosystems down to the microscale interactions within individual microhabitats. Each level of study provides unique insights into ecological patterns and processes.

6.4. BY LEVEL OF ORGANISATION OR SCOPE

Autecology

Autecology, also known as autecological ecology, is the subdiscipline of ecology that focuses on the study of individual species and their interactions with their environment. It is concerned with understanding the unique ecological requirements, adaptations, and behaviors of specific species.

Key Areas of Study:

A. Species-specific Ecology: Investigating the habitat preferences, life history traits, and ecological niches of individual species.

B. Population Dynamics: Examining how environmental factors and interactions influence the abundance and distribution of a particular species.

C. Species Conservation: Studying the specific needs and threats to individual species and devising conservation strategies.

D. Examples of Research Topics: Understanding the habitat requirements of a rare orchid species, studying the foraging behavior of a specific bird species, and assessing the conservation status of a critically endangered turtle species.

Synecology

Synecology, also known as synecological ecology, is the subdiscipline of ecology that focuses on the study of ecological communities and the interactions among different species within a defined area or habitat. It seeks to understand the composition, structure, and dynamics of communities.

Key Areas of Study:

A. Community Composition: Analyzing the species composition, diversity, and relative abundance within ecological communities.

B. Species Interactions: Investigating the ecological relationships among different species, including competition, predation, mutualism, and parasitism.

C. Succession and Community Change: Studying how communities change over time in response to ecological disturbances and environmental gradients.

D. Examples of Research Topics: Examining the food web dynamics in a freshwater lake ecosystem, studying the impact of invasive species on native plant communities, and assessing the factors influencing community composition in a tropical rainforest.

Population Ecology

Population ecology is the subdiscipline of ecology that focuses on the study of populations of the same species and their interactions within a given geographic area. It aims to understand population dynamics, growth, and regulation.

Key Areas of Study:

A. Population Growth Models: Investigating the factors influencing population growth, including birth rates, death rates, immigration, and emigration.

B. Density-Dependent and Density-Independent Factors: Analyzing how population size is influenced by factors such as resource availability, predation, and environmental conditions.

C. Population Conservation: Studying strategies for the conservation and management of endangered or threatened populations.

D. Examples of Research Topics: Assessing the impact of hunting on a local deer population, modeling the population dynamics of a fish species in a river, and developing conservation plans for a declining butterfly species.

Community Ecology

Community ecology is the subdiscipline of ecology that focuses on the study of ecological communities, including the interactions among different species and the factors that shape community composition and structure.

Key Areas of Study:

A. Species Interactions: Investigating the various types of interactions among species, such as predation, competition, facilitation, and symbiosis.

B. Species Diversity: Analyzing the patterns of species richness and diversity within communities and the factors influencing these patterns.

C. Trophic Relationships: Studying the flow of energy and matter through food webs and the role of keystone species in shaping communities.

D. Examples of Research Topics: Examining the impact of invasive species on native plant communities, studying the structure of a coral reef community, and analyzing the role of predators in controlling herbivore populations in a grassland ecosystem.

Ecosystem Ecology

Ecosystem ecology is the subdiscipline of ecology that focuses on the study of entire ecosystems, including the interactions among biotic and abiotic components, energy flow, nutrient cycling, and ecosystem processes.

Key Areas of Study:

A. Energy Flow: Investigating how energy enters, flows through, and exits ecosystems, including primary production and trophic dynamics.

B. Nutrient Cycling: Analyzing the cycling of essential nutrients (e.g., carbon, nitrogen, phosphorus) within ecosystems and their impact on ecosystem productivity.

C. Ecosystem Services: Studying the benefits provided by ecosystems to humans, such as clean water, carbon sequestration, and pollination.

D. Examples of Research Topics: Assessing the carbon storage capacity of a forest ecosystem, studying the nutrient cycling in a wetland ecosystem, and quantifying the ecosystem services provided by a coastal mangrove forest.

These classifications based on the level of organization or scope reflect the various ecological research approaches, from the study of individual species to the examination of entire ecosystems and their components. Ecologists in these subdisciplines contribute to a holistic understanding of ecological systems and their functioning.

6.5. BY BIOLOGICAL CLASSIFICATION OR TAXON UNDER STUDY

Human Ecology

Human ecology is the subdiscipline of ecology that focuses on the study of the interactions between humans and their environment. It explores how human activities, societies, and cultures influence and are influenced by ecological systems.

Key Areas of Study:

A. Anthropogenic Impact: Investigating the environmental impact of human activities, including deforestation, urbanization, and resource consumption.

B. Sustainable Development: Studying strategies for balancing human development with environmental conservation and the sustainable use of resources.

C. Human Health and Environment: Analyzing the connections between environmental factors and human health, including the spread of diseases and exposure to pollutants.

D. Examples of Research Topics: Assessing the environmental impact of agriculture and land use changes, studying the cultural practices of indigenous communities in relation to their ecosystems, analyzing the ecological consequences of urbanization.

Animal Ecology

Animal ecology is the subdiscipline of ecology that focuses on the study of animal species, their behaviors, population dynamics, and interactions with other organisms and their environment.

Key Areas of Study:

A. Behavioral Ecology: Investigating the behavior of animals, including mating, foraging, and communication.

B. Population Dynamics: Analyzing the abundance, distribution, and changes in animal populations.

C. Community Interactions: Studying how different animal species interact within ecosystems, including competition, predation, and mutualism.

D. Examples of Research Topics: Tracking the movement of migratory birds, studying the foraging behavior of wolves, and assessing the impact of human activities on wildlife populations.

Insect Ecology

Insect ecology is a specialized branch of ecology that focuses on the study of insects (arthropods) and their interactions with their environment.

Key Areas of Study:

A. Insect Behaviour: Investigating the behavior of insects, including mating, feeding, and communication.

B. Insect-Plant Interactions: Analyzing the relationships between insects and the plants they interact with, including herbivory and pollination.

C. Insect Population Dynamics: Studying the abundance and distribution of insect populations and the factors that influence them.

D. Examples of Research Topics: Examining the ecological role of pollinators in plant reproduction, studying the impact of invasive insect species on ecosystems, and researching the behavior of social insects like ants and bees.

Microbial Ecology

Microbial ecology is the subdiscipline of ecology that focuses on the study of microorganisms, including bacteria, fungi, archaea, and viruses, and their interactions with each other and their environments.

Key Areas of Study:

A. Microbial Communities: Investigating the composition and functions of microbial communities in various ecosystems, such as soils, oceans, and human microbiomes.

B. Nutrient Cycling: Analyzing the role of microorganisms in nutrient cycling, decomposition, and biogeochemical processes.

C. Microbial Interactions: Studying how microorganisms interact with each other and with higher organisms, including pathogenic interactions and symbiotic relationships.

D. Examples of Research Topics: Assessing the role of soil microbes in carbon sequestration, studying the microbial diversity in extreme environments like hydrothermal vents, and investigating the human gut microbiome and its impact on health.

Plant Ecology

Plant ecology is the subdiscipline of ecology that focuses on the study of plants and their interactions with their environment.

Key Areas of Study:

A. Vegetation Dynamics: Analyzing the composition and structure of plant communities and how they change over time.

B. Plant Adaptations: Investigating how plants have evolved to survive and thrive in various environmental conditions.

C. Ecosystem Function: Examining the role of plants in ecosystem processes, including photosynthesis, nutrient cycling, and carbon sequestration.

D. Examples of Research Topics: Studying the effects of climate change on plant distributions, researching the role of specific plant species in nutrient cycling, and restoring native vegetation in disturbed habitats.

These classifications based on the biological classification or taxon under study highlight the ecological study of different organisms and their relationships with their environments. Each subdiscipline contributes to our understanding of ecological processes and biodiversity within specific biological groups.

6.6. BY BIOME UNDER STUDY

Figure 40. Biomes

(Source: activewild, Creative Commons License)

Benthic Ecology

Benthic ecology is the subdiscipline of ecology that focuses on the study of organisms and ecological processes in benthic environments, which include the bottoms of aquatic ecosystems such as oceans, lakes, rivers, and estuaries.

Key Areas of Study:

A. Benthic Communities: Investigating the composition and diversity of organisms living on or in the sediment or substrate.

B. Sediment Dynamics: Analyzing the physical and chemical properties of benthic substrates and their influence on organism behaviour and distribution.

C. Benthic Food Webs: Studying the flow of energy and nutrients among benthic organisms and their relationships with pelagic (open water) ecosystems.

D. Examples of Research Topics: Assessing the impact of bottom trawling on benthic communities, studying the biodiversity of deep-sea hydrothermal vents, and analyzing the role of benthic organisms in nutrient cycling in estuaries.

Desert Ecology

Desert ecology is the subdiscipline of ecology that focuses on the study of desert ecosystems, characterized by arid conditions, low precipitation, and unique adaptations of organisms to survive in extreme desert environments.

Key Areas of Study:

A. Desert Flora and Fauna: Investigating the plant and animal species that have adapted to desert conditions.

B. Water and Resource Management: Analyzing strategies for water conservation and resource utilization by desert organisms.

C. Desertification and Climate Change: Studying the impacts of desertification processes and climate change on desert ecosystems.

D. Examples of Research Topics: Assessing the ecological roles of desert keystone species, studying the water-saving adaptations of desert plants, and analyzing the effects of climate change on desert wildlife.

Forest Ecology

Forest ecology is the subdiscipline of ecology that focuses on the study of forest ecosystems, including temperate and tropical forests, and the interactions among the organisms and processes within these ecosystems.

Key Areas of Study:

A. Forest Structure and Composition: Investigating the types of trees, understory vegetation, and wildlife in forest ecosystems.

B. Nutrient Cycling: Analyzing the cycling of nutrients, such as carbon, nitrogen, and phosphorus, within forest ecosystems.

C. Forest Management and Conservation: Studying sustainable forestry practices, habitat conservation, and the impact of deforestation.

D. Examples of Research Topics: Assessing the impact of logging on forest biodiversity, studying the role of old-growth forests in carbon sequestration, and analyzing the effects of invasive species on temperate forest ecosystems.

Grassland Ecology

Grassland ecology is the subdiscipline of ecology that focuses on the study of grassland ecosystems, characterized by open landscapes dominated by grasses and herbaceous plants.

Key Areas of Study:

A. Grassland Biodiversity: Investigating the species diversity and composition of grassland communities.

B. Grazing and Fire Ecology: Analyzing the ecological role of grazing animals and natural fire regimes in grasslands.

C. Grassland Restoration: Studying efforts to restore and conserve grassland ecosystems, including the reintroduction of native species.

D. Examples of Research Topics: Assessing the impact of overgrazing on grassland plant diversity, studying the role of fire in maintaining grassland habitats, and analyzing the effects of agricultural practices on grassland bird populations.

Marine Ecology

Marine ecology is the subdiscipline of ecology that focuses on the study of marine ecosystems, including oceans, seas, and coastal environments, and the interactions among marine organisms and their surroundings.

Key Areas of Study:

A. Coral Reef Ecology: Investigating the biodiversity, conservation, and health of coral reef ecosystems.

B. Pelagic Ecosystems: Analyzing open ocean ecosystems and the dynamics of marine food webs.

C. Marine Conservation: Studying the impacts of overfishing, pollution, and climate change on marine biodiversity.

D. Examples of Research Topics: Assessing the effects of ocean acidification on coral reefs, studying the migration patterns of marine mammals, and analyzing the role of marine protected areas in conserving marine life.

Aquatic Ecology

Aquatic ecology is the subdiscipline of ecology that encompasses the study of freshwater ecosystems, including lakes, rivers, wetlands, and streams, and the interactions among aquatic organisms and their environments.

Key Areas of Study:

A. Freshwater Biodiversity: Investigating the diversity and distribution of aquatic species in freshwater habitats.

B. Aquatic Food Webs: Analyzing the flow of energy and nutrients in aquatic ecosystems.

C. Wetland Ecology: Studying the ecological functions and conservation of wetland habitats.

D. Examples of Research Topics: Assessing the impact of nutrient pollution on lake eutrophication, studying the migration of fish in river systems, and analyzing the role of wetlands in water purification.

Urban Ecology

Urban ecology is the subdiscipline of ecology that focuses on the study of ecosystems within urban environments, including cities and metropolitan areas, and the ecological processes in these human-modified landscapes.

Key Areas of Study:

A. Urban Biodiversity: Investigating the presence and diversity of species in urban settings.
B. Urban Green Spaces: Analyzing the ecological value and functions of urban parks, gardens, and green infrastructure.
C. Urban Sustainability: Studying strategies for sustainable urban development, green building, and resource management.
D. Examples of Research Topics: Assessing the impact of urbanization on bird diversity, studying the ecological benefits of rooftop gardens, and analyzing urban heat island effects on local climate.

These classifications based on the biome or ecosystem type under study highlight the diversity of ecological research within different terrestrial and aquatic environments, from deserts and forests to urban landscapes and aquatic ecosystems. Each subdiscipline contributes to our understanding of the unique challenges and ecological dynamics of these environments.

6.7. BY BIOGEOGRAPHIC REALM OR CLIMATIC AREA UNDER STUDY

Arctic Ecology

Arctic ecology is the subdiscipline of ecology that focuses on the study of ecological systems and processes in the Arctic region, characterized by extreme cold, permafrost, and unique adaptations of organisms to Arctic conditions.

Key Areas of Study:

- Cold-Adapted Species: Investigating the specialized adaptations of Arctic organisms to survive in freezing temperatures.
- Climate Change Impacts: Analyzing the effects of climate change on Arctic ecosystems, including melting sea ice and shifts in species distributions.
- Arctic Conservation: Studying efforts to conserve Arctic biodiversity and protect fragile Arctic habitats.
- Examples of Research Topics: Assessing the impact of warming temperatures on Arctic tundra vegetation, studying the ecology of polar bears in relation to sea ice loss, and analyzing the role of

Arctic ecosystems in global carbon cycling.

Polar Ecology:

Polar ecology is the subdiscipline of ecology that encompasses the study of ecosystems and ecological processes in polar regions, including both the Arctic and Antarctic, characterized by extreme cold, ice-covered landscapes, and unique polar adaptations of organisms.

Key Areas of Study:

- Polar Marine Ecosystems: Investigating the biodiversity and ecological dynamics of polar oceans and sea ice.
- Polar Terrestrial Ecosystems: Analyzing the flora and fauna of polar terrestrial environments, including mosses, lichens, and cold-adapted animals.
- Polar Food Webs: Studying the flow of energy and nutrients in polar ecosystems and the role of key species like krill and seals.
- Examples of Research Topics: Assessing the impact of climate change on Antarctic ice shelves and wildlife, studying the adaptations of penguins to harsh polar conditions, and analyzing the ecological role of Arctic sea ice algae.

Tropical Ecology:

Tropical ecology is the subdiscipline of ecology that focuses on the study of ecosystems in tropical regions, characterized by high biodiversity, warm temperatures, and diverse ecological interactions.

Key Areas of Study:

- Rainforest Ecology: Investigating the structure, diversity, and ecological functions of tropical rainforests.
- Coral Reef Ecology: Analyzing the biodiversity and conservation of coral reef ecosystems.
- Tropical Biodiversity: Studying the unique flora and fauna of tropical regions and their adaptations.
- Examples of Research Topics: Assessing the impact of deforestation on Amazon rainforest biodiversity, studying the resilience of coral reefs to bleaching events, and analyzing the ecological role of keystone species in tropical ecosystems.

These classifications based on the biogeographic realm or climatic area reflect the diverse ecological research conducted in extreme cold regions, polar environments, and lush tropical landscapes. Each subdiscipline contributes to our understanding of the ecological dynamics and conservation challenges specific to these areas.

6.8. BY ECOLOGICAL ASPECTS OR PHENOMENA UNDER INVESTIGATION

Behavioral Ecology

Behavioral ecology is the subdiscipline of ecology that focuses on the study of animal behaviors in the context of their ecological roles, including mating, foraging, predator-prey interactions, communication, and social behaviors.

Key Areas of Study:

A. Investigating the adaptive significance of behaviors, the influence of ecological factors on behavior, and the fitness consequences of behavioral choices.

B. Examples of Research Topics: Studying the mating behavior of birds, analyzing foraging strategies of predators, and investigating the communication behaviors of social insects.

Chemical Ecology

Chemical ecology is the subdiscipline of ecology that explores the chemical interactions among organisms and their environments. It focuses on chemical signals, allelopathy, and the role of chemical compounds in ecological interactions.

Key Areas of Study:

A. Investigating chemical cucs used in predator-prey relationships, chemical defenses in plants, and the chemical basis of mutualistic interactions.

B. Examples of Research Topics: Studying pheromones in insect mating, analyzing allelopathic compounds in plant competition, and exploring chemical defenses in marine organisms.

Disease Ecology

Disease ecology is the subdiscipline of ecology that examines the dynamics of infectious diseases within populations and ecosystems, including the spread of pathogens, host-pathogen interactions, and disease impacts on populations.

Key Areas of Study:

A. Investigating disease transmission, the ecological factors influencing disease outbreaks, and the role of hosts and vectors in disease cycles.

B. Examples of Research Topics: Studying the spread of vector-borne diseases like malaria, analyzing the impact of diseases on wildlife populations, and investigating disease emergence in changing ecosystems.

Ecophysiology

Ecophysiology is the subdiscipline of ecology that focuses on the physiological adaptations of organisms to their environments, including how physiological traits affect an organism's ecological niche.

Key Areas of Study:

A. Investigating how organisms regulate temperature, water balance, and metabolic processes in response to environmental conditions.

B. Examples of Research Topics: Studying how animals in hot deserts thermoregulate, analyzing the water use efficiency of plants in arid environments, and examining the metabolic adaptations of high-altitude species.

Ecotoxicology

Ecotoxicology is the subdiscipline of ecology that explores the effects of pollutants and contaminants on ecosystems and organisms. It focuses on the impact of toxic substances on ecological processes and biodiversity.

Key Areas of Study:

A. Investigating the toxicity of chemicals, biomagnification of pollutants in food webs, and the ecological risks of pollution.

B. Examples of Research Topics: Studying the effects of pesticides on aquatic ecosystems, analyzing the impact of heavy metal contamination on soil organisms, and assessing the ecological consequences of oil spills.

Evolutionary Ecology

Evolutionary ecology is the subdiscipline of ecology that examines how ecological processes and interactions drive evolutionary changes in populations and species.

Key Areas of Study:

A. Investigating the role of natural selection, adaptation, and coevolution in shaping ecological communities.

B. Examples of Research Topics: Studying the evolution of predator-prey interactions, analyzing the coevolution of plants and their pollinators, and investigating the adaptation of species to changing environments.

Functional Ecology

Functional ecology is the subdiscipline of ecology that focuses on the roles and functions of organisms in ecosystems, including how species traits and behaviors influence ecosystem processes.

Key Areas of Study:

A. Investigating how species traits affect ecosystem functions, such as nutrient cycling, decomposition, and primary productivity.

B. Examples of Research Topics: Studying the functional diversity of plant species in grasslands, analyzing the role of herbivores in nutrient cycling, and investigating the functional traits of aquatic microbes.

Genetic Ecology

Genetic ecology is the subdiscipline of ecology that examines how genetic variation within populations influences ecological processes, including adaptation, dispersal, and response to environmental change.

Key Areas of Study:

A. Investigating the genetic basis of traits, gene flow among populations, and the impact of genetic diversity on ecosystem resilience.

B. Examples of Research Topics: Studying the genetic diversity of fish populations in river systems, analyzing the genetic adaptations of plants to different soil conditions, and assessing the genetic diversity of coral populations.

Landscape Ecology and Limnology

Landscape ecology focuses on the spatial arrangement of ecosystems and the connectivity between them, while limnology is the study of freshwater ecosystems, including lakes, rivers, and wetlands.

Key Areas of Study:

A. Investigating landscape patterns, habitat fragmentation, hydrology, and the ecological dynamics of freshwater ecosystems.

B. Examples of Research Topics: Studying the impact of habitat fragmentation on biodiversity, analyzing the nutrient cycling in freshwater lakes, and assessing the connectivity of aquatic habitats in river systems.

Molecular Ecology

Molecular ecology is the subdiscipline of ecology that uses molecular genetic techniques to study ecological questions, including genetic diversity, population structure, and gene-environment interactions.

Key Areas of Study:

A. Investigating DNA markers, gene expression, and genomics in ecological research.

B. Examples of Research Topics: Studying the genetic diversity of endangered species, analyzing gene flow in plant populations, and using DNA barcoding to identify species in ecological communities.

Paleoecology

Paleoecology is the subdiscipline of ecology that reconstructs past ecosystems and environmental conditions using fossils, sediment cores, and other historical data.

Key Areas of Study:

A. Investigating ancient climates, ecological changes over geological time scales, and the responses of past ecosystems to environmental shifts.

B. Examples of Research Topics: Studying the impact of climate change on past megafauna extinctions, reconstructing ancient ecosystems from fossil records, and analyzing the historical distribution of plant species.

Social Ecology

Social ecology is the subdiscipline of ecology that focuses on the interactions between humans and the environment, including the social, cultural, and economic aspects of environmental issues.

Key Areas of Study:

A. Investigating human-environment interactions, environmental policy, sustainability, and the social dimensions of conservation.

B. Examples of Research Topics: Studying the impact of land-use policies on ecosystem services, analyzing community-based conservation approaches, and assessing the environmental implications of urbanization.

Soil Ecology

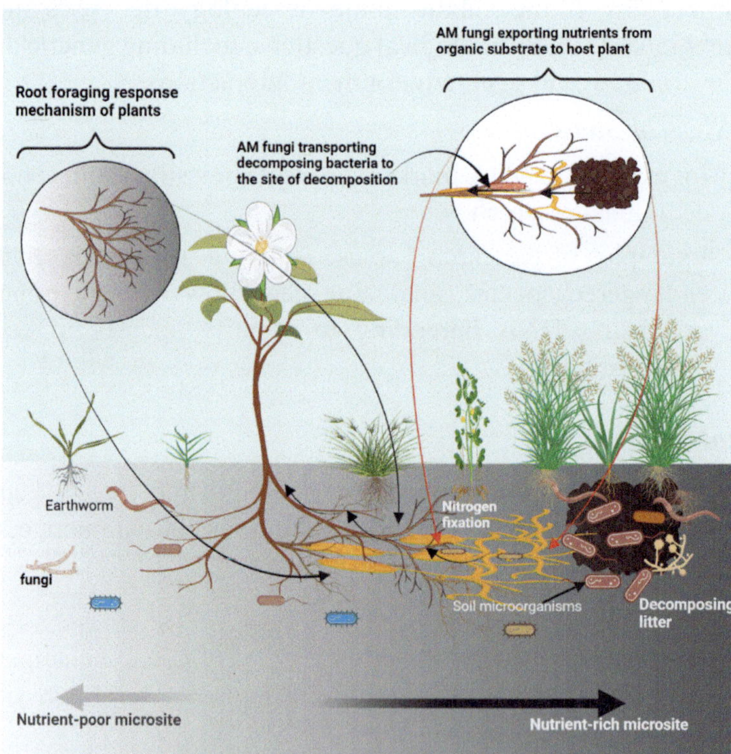

Figure 41. Soil Ecology

(Source: Adomako, M. O., Roiloa, S., Yu, F.-H. Potential Roles of Soil Microorganisms in Regulating the Effect of Soil Nutrient Heterogeneity on Plant Performance. Microorganisms. 2022; 10(12):2399. https://doi.org/10.3390/microorganisms10122399, Creative Commons License)

Soil ecology is the subdiscipline of ecology that studies the interactions among organisms in the soil ecosystem, including microorganisms, plants, and invertebrates, and their influence on soil health and fertility.

Key Areas of Study:

A. Investigating nutrient cycling, soil structure, microbial communities, and the role of soil organisms in ecosystem processes.

B. Examples of Research Topics: Studying the effects of soil microorganisms on plant nutrient uptake, analyzing the impact of land management practices on soil biodiversity, and assessing the role of earthworms in soil aeration and nutrient cycling.

Spatial Ecology

Spatial ecology is the subdiscipline of ecology that focuses on the analysis of spatial patterns and processes in ecological systems, including habitat distribution, movement, and dispersal of organisms.

Key Areas of Study:

A. Investigating habitat fragmentation, connectivity, spatial modeling, and the movement of species across landscapes.

B. Examples of Research Topics: Studying the impact of roads on wildlife movement, analyzing the spatial distribution of rare species, and assessing the effectiveness of protected areas in conserving biodiversity.

Thermal Ecology:

Thermal ecology is the subdiscipline of ecology that explores the relationships between organisms and temperature, including how temperature affects physiological processes, behavior, and species distributions.

Key Areas of Study:

A. Investigating temperature adaptations, thermal tolerances, and the effects of climate change on thermal regimes.

B. Examples of Research Topics: Studying the heat tolerance of desert animals, analyzing the impact of warming temperatures on alpine plant communities, and assessing the thermal preferences of aquatic species.

The aforementioned classifications based on ecological aspects or phenomena under investigation demonstrate the breadth of ecological research, from understanding animal behavior to exploring the molecular genetics of species and the impact of human activities on ecosystems. Each subdiscipline contributes to our understanding of different aspects of ecological science.

6.9. REFERENCES

1. Allee, W. C. (1932). Animal Life and Social Growth. Baltimore: The Williams & Wilkins Company and Associates.
2. Allee, W., Emerson, A. E., Park, O., Park, T., & Schmidt, K. P. (1949). Principles of Animal Ecology. W. B. Saunders Company. ISBN 0721611206.
3. Begon, M., Townsend, C. R., & Harper, J. L. (2006). Ecology: From Individuals to Ecosystems. (4th ed.). Blackwell. ISBN 1405111178.
4. Brinson, M. M., Lugo, A. E., & Brown, S. (1984). "Primary Productivity, Decomposition and Consumer Activity in Freshwater Wetlands." Annual Review of Ecology and Systematics 12, 123–161. doi:10.1146/annurev.es.12.110181.001011. http://www.jstor, (1905). Research Methods in Ecology. Lincoln, Nebraska: University Publ. ISBN 0405103816.
5. Costanza, R., d'Arge, R., de Groot, R., Farberk, S., Grasso, M., Hannon, B., et al., (1997). "The value of the world's ecosystem services and natural capital." Nature 387, 253–260. doi:10.1038/387253a0. http://www.uvm.edu/giee/publications/Nature_Paper.pdf.
6. Davic, R. D., & Welsh, H. H. (2004). "On the Ecological Role of Salamanders." Annual Review of Ecology and Systematics 35, 405–434. doi:10.1146/annurev.ecolsys.35.112202.130116.
7. Elton, C. S. (1927). Animal Ecology. London, UK.: Sidgwick and Jackson. ISBN 0226206394.
8. Forbes, S. (1887). "The lake as a microcosm." Bull. of the Scientific Association (Peoria, IL): 77–87. http://www.uam.es/personal_pdi/ciencias/scasado/documentos/Forbes.PDF.
9. Gleason, H. A. (1926). "The Individualistic Concept of the Plant Association." Bulletin of the Torrey Botanical Club 53(1), 7–26. doi:10.2307/2479933. http://www.ecologia.unam.mx/laboratorios/comunidades/pdf/pdf%20curso%20posgrado%20Elena/Tema%201/gleason1926.pdf.
10. Hanski, I. (2008). "The world that became ruined." EMBO Reports 9, S34–S36. doi:10.1038/embor.2008.77. PMID 18578021. http://www.helsinki.fi/~ihanski/Articles/EMBO%202008%20Hanski.pdf.
11. Hastings, A. B., Crooks, J. E., Cuddington, J. A., Jones, K., Lambrinos, C. J., Talley, J. G., et al., TS; Wilson, WG (2007). "Ecosystem

engineering in space and time." Ecology Letters 10(2): 153–164. doi:10.1111/j.1461–0248.2006.00997.10. PMID 17257103.

12. Kormondy, E. (1995). Concepts of ecology. (4th ed.). Benjamin Cummings. ISBN 0134781163.
13. Liu, J., Dietz, T., Carpenter, S. R., Folke, C., Alberti, M., Redman, C. L., et al., Stephen H., Ostrom, Elinor et al. (2009). "Coupled Human and Natural Systems." AMBIO: A Journal of the Human Environment 36(8), 639–649. doi:10.1579/0044–7447(2007)36[639:CHANS]2.0.CO;2. http://ambio.allenpress, Creative Commons License).

CHAPTER 7

Interdisciplinary Ecology

CONTENTS

7.1. INTRODUCTION TO INTERDISCIPLINARY ECOLOGY

Ecology, the scientific study of the relationships between living organisms and their environments, forms the foundation of our understanding of the natural world. It delves into the intricate web of life, exploring how species interact, how ecosystems function, and how environmental factors shape the diversity of life on Earth. However, as ecological challenges become increasingly complex and intertwined with social, economic, and technological factors, the need for interdisciplinary approaches in ecological research has never been more crucial. This introduction will illuminate the significance of interdisciplinary approaches in ecology and elucidate the multifaceted web that connects ecology with related scientific disciplines.

Interdisciplinary ecology, often referred to as the fusion of ecological knowledge with insights from various scientific realms, is indispensable in addressing the intricate environmental issues of our time. Several factors underscore the importance of interdisciplinary approaches in ecology:

- Complexity of Natural Systems: Ecological systems are intricate and multifaceted, involving diverse species, ecological processes, and environmental factors. Understanding and managing such complexity necessitates the integration of knowledge from various scientific domains.
- Global Environmental Challenges: Issues like climate change, habitat degradation, loss of biodiversity, and pollution transcend disciplinary boundaries. Interdisciplinary research provides a more comprehensive understanding of these challenges and the development of effective solutions.
- Human-Ecological Interactions: Human activities exert profound influences on ecosystems and natural resources. Interdisciplinary approaches are essential for assessing the impact of human actions on the environment and devising sustainable solutions.
- Innovation and Holistic Understanding: Collaboration across disciplines fosters innovation, encourages fresh perspectives, and promotes holistic comprehension. Complex ecological problems often require multifaceted solutions that draw upon diverse fields of knowledge.
- Sustainable Resource Management: Interdisciplinary research is vital for informed decision-making in resource management.

It assists in balancing ecological conservation with economic development and societal needs.

Understanding the Nexus of Ecology and Related Disciplines:

- Ecology's connection with other scientific disciplines is both intricate and profound. It spans a broad spectrum of domains, each contributing a unique perspective to our understanding of the natural world. Here are some key disciplines that intersect with ecology:
- Environmental Science: Environmental science embraces the holistic study of the environment, encompassing ecology, geology, chemistry, physics, and social sciences. It addresses environmental issues from a multidisciplinary viewpoint.
- Conservation Biology: Conservation biology combines ecological principles with genetics, sociology, economics, and policy to develop strategies for preserving biodiversity and protecting endangered species.
- Climate Science: Climate science investigates the Earth's climate system, including the ecological consequences of climate change. It collaborates with ecology to assess how shifting climate patterns impact ecosystems.
- Biogeography: Biogeography explores species distributions across geographical regions, merging ecology with geography, geology, and evolutionary biology to understand the factors influencing these distributions.
- Landscape Ecology: Landscape ecology studies the spatial patterns and arrangements of landscapes, linking ecological processes with geography and land-use planning. It considers ecological dynamics at various spatial scales.
- Human Ecology: Human ecology investigates the complex interactions between human societies and ecosystems. It explores the cultural, economic, and social dimensions of ecological issues.
- Socioecology: Socioecology delves into the social aspects of ecological systems, considering how human communities interact with and influence ecosystems. It examines the role of culture, economics, and social factors in shaping ecological outcomes.
- Ecohydrology: Ecohydrology examines the interactions between

ecological processes and water availability. It bridges ecology and hydrology to understand the role of water in ecosystems.

- Ethnobotany: Ethnobotany combines ecological knowledge with cultural insights to understand how indigenous communities use and conserve plant resources.
- Ecoinformatics: Ecoinformatics applies computational tools and data analysis to ecological research, aiding in deciphering complex ecological patterns and relationships.

The connections between ecology and these related disciplines underscores the interplay between ecological systems and the broader scientific, societal, and economic contexts in which they exist. Interdisciplinary approaches empower us to navigate the complexities of the modern world, providing a more comprehensive and integrated perspective on ecological challenges and opportunities. In the pages that follow, we will delve deeper into the realms of interdisciplinary ecology, exploring how these collaborations contribute to a more profound understanding of our planet's ecosystems and the sustainable coexistence of all life forms.

7.2. AGROECOLOGY

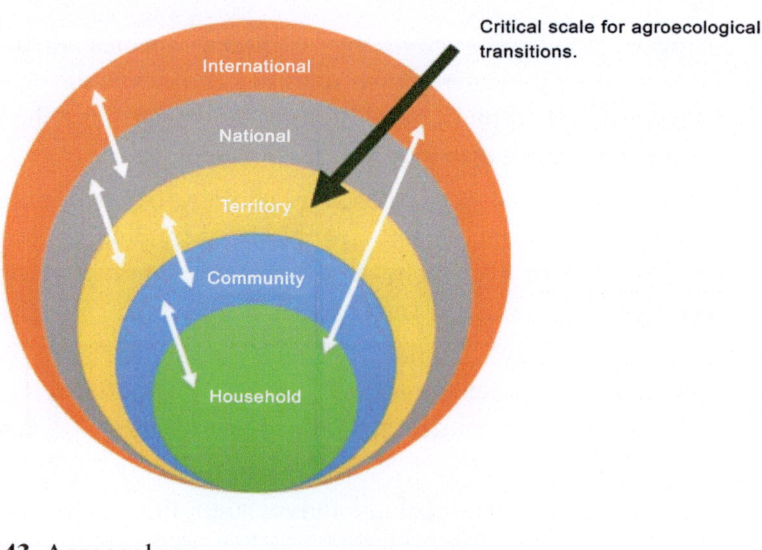

Figure 43. Agroecology

(Source: Anderson, C. R., Bruil, J., Chappell, M. J., Kiss, C., Pimbert, M. P. From Transition to Domains of Transformation: Getting to Sustainable and Just

Food Systems through Agroecology. Sustainability. 2019; 11(19), 5272. https://doi.org/10.3390/su11195272, Creative Commons License)

Agroecology is a holistic and interdisciplinary approach to agriculture that not only addresses the production of food but also the ecological, social, and economic dimensions of farming. It emphasizes the interdependence of agricultural practices with the environment and seeks to enhance sustainability, resilience, and food security. In this exploration of agroecology, we will delve into its principles, its role in promoting sustainable agriculture and food systems, and the agroecological practices that contribute to environmental conservation.

Principles of Agroecology

- Biodiversity: Agroecology recognizes the importance of biodiversity in agricultural systems. Diverse ecosystems are more resilient to pests, diseases, and environmental stresses. Polyculture (growing multiple crops together) and integrating livestock with crops are practices that enhance biodiversity.
- Ecosystem Services: Agroecological systems aim to mimic natural ecosystems, leveraging ecosystem services such as pollination, nutrient cycling, and pest control. By working with nature rather than against it, agroecology reduces the need for synthetic inputs.
- Sustainability: Sustainable practices are at the core of agroecology. It prioritizes long-term environmental and social sustainability, avoiding practices that deplete natural resources or harm ecosystems. Crop rotation, organic farming, and reduced tillage are examples of sustainable practices.
- Local Adaptation: Agroecology promotes locally adapted solutions to agricultural challenges. It recognizes that the best practices for one region may not be suitable for another. Farmers are encouraged to develop context-specific approaches.
- Soil Health: Soil is a foundational component of agriculture. Agroecology emphasizes soil health and fertility through practices like cover cropping, composting, and reduced chemical inputs. Healthy soil is essential for crop productivity and carbon sequestration.

Sustainable Agriculture and Food Systems

Agroecology plays a pivotal role in fostering sustainable agriculture and food systems:

- Diverse and Resilient Farms: Agroecological practices encourage diversified farms that grow a variety of crops and incorporate animals. This diversity enhances resilience to pests, diseases, and climatic variations.
- Reduced Chemical Inputs: Agroecology reduces the reliance on synthetic pesticides and fertilizers, which can harm the environment and human health. Organic farming methods are often embraced.
- Food Security: By emphasizing local production and diverse crops, agroecology contributes to food security. It helps small-scale farmers maintain autonomy over their food production and distribution.
- Conservation of Genetic Diversity: Agroecology promotes the conservation of traditional and locally adapted crop varieties, safeguarding genetic diversity for future food security.
- Climate Resilience: Agroecological practices help farmers adapt to and mitigate climate change. Practices like agroforestry and water management contribute to climate resilience and carbon sequestration.

Agroecological Practices for Environmental Conservation:

Agroecology's environmental conservation practices are instrumental in protecting ecosystems and natural resources:

- Agroforestry: Integrating trees and shrubs into agricultural landscapcs cnhances biodiversity, provides shade, sequesters carbon, and reduces soil erosion.
- Crop Rotation: Crop rotation reduces soil depletion, prevents the build-up of pests and diseases, and enhances soil fertility.
- Cover Cropping: Planting cover crops during fallow periods reduces soil erosion, improves soil structure, and adds organic matter.
- Polyculture: Growing multiple crops together simulates natural ecosystems and enhances pest control through diversity.

- Integrated Pest Management (IPM): IPM strategies in agroecology emphasize biological control methods, like introducing natural predators to manage pests.
- Water Management: Sustainable water management practices, such as rainwater harvesting and drip irrigation, minimize water wastage and reduce the pressure on freshwater resources.
- Composting and Organic Matter: Adding compost and organic matter to soil improves its water-holding capacity and nutrient content while reducing the need for synthetic fertilizers.

Agroecology embodies a profound shift in agricultural paradigms, emphasizing that farming can and should be aligned with the principles of ecological sustainability and environmental conservation. It recognizes that agriculture is not separate from nature but deeply interconnected with it. As we delve deeper into the realm of agroecology, we will explore the real-world applications of these principles and practices, demonstrating how they contribute not only to sustainable agriculture but also to the preservation of our planet's ecosystems and the well-being of future generations.

7.3. APPLIED ECOLOGY

Applied ecology represents the practical application of ecological principles and concepts to address real-world environmental challenges. It serves as a critical link between ecological research and tangible solutions for pressing environmental issues. In this comprehensive exploration of applied ecology, we will delve into the practical applications of ecological concepts, the pivotal role of applied ecology in environmental problem-solving, and its specific branch, conservation ecology, as well as restoration ecology.

Practical Applications of Ecological Concepts:

Natural Resource Management: Applied ecology guides the sustainable use of natural resources such as forests, fisheries, and water bodies. It informs practices that balance resource extraction with conservation.

Agriculture and Land Use: Ecological principles are applied to optimize agricultural practices, reduce soil erosion, and enhance crop yield while minimizing environmental impacts.

Urban Planning: Applied ecology plays a role in designing ecologically sustainable cities by addressing issues like green infrastructure, urban biodiversity, and waste management. Ecosystem Health Assessment: Ecological concepts are used to evaluate the health of ecosystems, detect pollution, and identify indicators of environmental degradation.

Role of Applied Ecology in Environmental Problem-Solving:

Environmental Impact Assessment (EIA): Applied ecologists contribute to EIA processes by evaluating the potential ecological consequences of development projects, helping minimize adverse effects on the environment.

Mitigation and Restoration: Applied ecology provides strategies for mitigating the impact of human activities, such as habitat restoration and creating wildlife corridors to compensate for habitat loss.

Conservation Planning: It aids in the identification of critical habitats, conservation priorities, and the development of protected area networks to safeguard biodiversity.

Conservation Ecology:

Biodiversity Conservation Strategies: Applied ecology informs strategies to protect and restore biodiversity, including:

- Protected Areas: Designing and managing protected areas, such as national parks and reserves, to conserve ecosystems and species.
- Habitat Connectivity: Creating corridors that link fragmented habitats to facilitate species movement and gene flow.
- Community-Based Conservation: Engaging local communities in conservation efforts and sustainable resource management.
- Conservation of Threatened and Endangered Species: Applied ecology plays a central role in the conservation of endangered species by:
- Population Monitoring: Tracking populations, understanding their ecological requirements, and implementing recovery plans.
- Ex Situ Conservation: Establishing captive breeding programs and seed banks to ensure genetic diversity.
- Reintroduction Programs: Releasing captive-bred or rehabilitated animals back into their natural habitats.
- Habitat Restoration and Protection: Applied ecology contributes to the restoration and protection of degraded habitats through:
- Wetland Restoration: Restoring wetlands to improve water

quality, enhance biodiversity, and mitigate flood risks.

- Riparian Zone Protection: Protecting riparian zones along rivers and streams to reduce erosion and provide habitat for aquatic life.
- Reforestation and Afforestation: Planting trees to restore deforested or degraded landscapes and sequester carbon.

Restoration Ecology

- Ecological Restoration Principles: Restoration ecology involves the application of ecological principles to the recovery of ecosystems:
- Set Clear Goals: Establish specific restoration objectives, such as restoring a wetland to support native bird species.
- Identify Reference Ecosystems: Use undisturbed or minimally disturbed ecosystems as models for restoration.
- Monitor and Adapt: Continuously assess progress and adjust restoration efforts based on results.
- Engage Stakeholders: Involve local communities and stakeholders in decision-making and restoration activities.
- Ecosystem Recovery and Rehabilitation: Restoration ecology focuses on the rehabilitation of ecosystems:
- Revegetation: Planting native species to restore vegetation cover and enhance soil stability.
- Soil Remediation: Treating contaminated soils to restore soil health.
- Invasive Species Control: Managing invasive species to reduce competition with native flora and fauna.
- Case Studies in Restoration Ecology: Real-world examples of restoration success stories, such as the restoration of the Florida Everglades or the rehabilitation of mining sites into biodiverse habitats, highlight the effectiveness of restoration ecology in recovering degraded ecosystems.

Applied ecology, encompassing conservation and restoration ecology, represents a bridge between ecological knowledge and practical environmental solutions. It offers hope for mitigating the ecological damage caused by human activities and working toward a more sustainable and harmonious coexistence with the natural world. As we delve further into these topics, we will uncover the innovative and inspiring ways in which applied ecology transforms ecological insights into tangible actions for a healthier planet.

7.4. BIOGEOCHEMISTRY

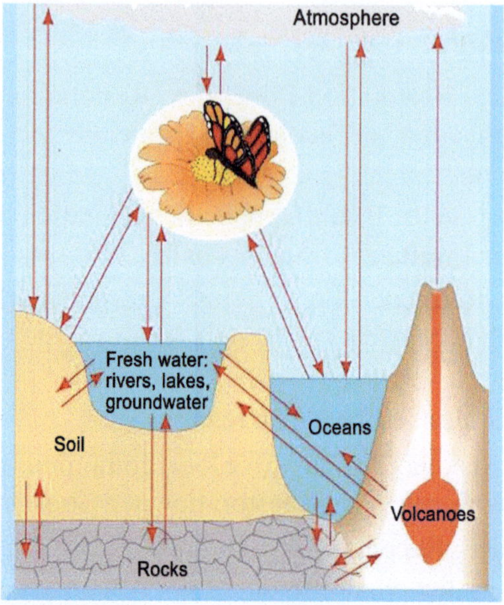

Figure 44. Biogeochemistry

(Source: By Muhumuza Moses – [1], CC BY-SA 3.0, https://commons.wikimedia.org/w/index.php?curid=110381197, Creative Commons License)

Biogeochemistry is the science that explores the interactions between living organisms, geological processes, and the cycling of chemical elements in ecosystems. It delves into the dynamic processes that govern the movement of essential elements like carbon, nitrogen, and phosphorus through the Earth's various spheres, including the atmosphere, biosphere, hydrosphere, and lithosphere. In this comprehensive examination of biogeochemistry, we will explore the cycling of elements in ecosystems, the significance of biogeochemical cycles, and the profound impacts of human activities on these essential processes.

The Cycling of Elements in Ecosystems

Elemental Exchange: Ecosystems continually exchange elements among living organisms, the atmosphere, water bodies, and the Earth's crust. This exchange is essential for life as it provides the raw materials for growth and metabolism.

Biological Processes: Living organisms play a pivotal role in biogeochemical cycles. They absorb elements from the environment, use them for growth and energy, and return them through processes like respiration, decomposition, and excretion.

Abiotic Processes: Geological processes, such as weathering and erosion, release elements from rocks and minerals, making them available for uptake by plants and microorganisms.

Chemical Transformations: Biogeochemical cycles involve chemical transformations of elements. For example, nitrogen undergoes various chemical reactions, including fixation, nitrification, and denitrification.

Biogeochemical Cycles: Carbon, Nitrogen, and Phosphorus

Carbon Cycle: The carbon cycle regulates the flow of carbon among the atmosphere, terrestrial ecosystems, and the oceans. Key processes include photosynthesis (removing carbon dioxide from the atmosphere), respiration (releasing carbon dioxide), and carbon storage in forests and soils. Human activities, such as burning fossil fuels and deforestation, have disrupted this cycle, leading to increased atmospheric carbon dioxide and contributing to climate change.

Nitrogen Cycle: The nitrogen cycle involves the transformation of nitrogen gas (N_2) in the atmosphere into forms that can be used by plants and animals. Nitrogen-fixing bacteria convert N_2 into ammonia (NH_3), which can be taken up by plants. Nitrogen cycling also includes nitrification (conversion of ammonia to nitrate) and denitrification (conversion of nitrate to N_2). Human activities, such as fertilizer use and industrial nitrogen fixation, have significantly altered the nitrogen cycle, leading to issues like nitrogen pollution and eutrophication.

Phosphorus Cycle: Phosphorus is essential for DNA, RNA, and ATP (adenosine triphosphate), crucial molecules for life. The phosphorus cycle primarily involves the weathering of rocks, phosphate uptake by plants, and the transfer of phosphorus through the food web. Excessive use of phosphorus-based fertilizers and improper waste disposal can lead to phosphorus runoff, causing water pollution and ecosystem disruptions.

Human Impacts on Biogeochemical Processes:

Alteration of the Carbon Cycle: The burning of fossil fuels and deforestation have released vast amounts of carbon dioxide into the atmosphere, contributing to global warming and climate change.

Nitrogen Pollution: Excessive use of nitrogen-based fertilizers and industrial nitrogen fixation have led to nitrogen pollution of water bodies, causing eutrophication, harmful algal blooms, and disruptions in aquatic ecosystems.

Phosphorus Runoff: Improper disposal of phosphorus-containing waste and agricultural runoff have resulted in phosphorus pollution in freshwater systems, leading to algal blooms and water quality deterioration.

- Alteration of Water and Nutrient Cycles: Urbanization and land-use changes have disrupted natural hydrological cycles, leading to increased runoff, soil erosion, and changes in nutrient availability.
- Deforestation and Habitat Loss: The removal of forests disrupts the carbon and nutrient cycles, reduces biodiversity, and alters the functioning of ecosystems.
- Ocean Acidification: Increased atmospheric carbon dioxide is leading to ocean acidification, which negatively impacts marine ecosystems and organisms with calcium carbonate shells or skeletons.

Understanding and mitigating the human impacts on biogeochemical processes are critical for maintaining the stability and health of ecosystems and addressing global environmental challenges such as climate change, nutrient pollution, and habitat loss. As we delve further into the realm of biogeochemistry, we will explore the complex interactions between living organisms, geological processes, and the Earth's elemental cycles, and how humanity can navigate these challenges for a sustainable future.

7.5. BIOGEOGRAPHY

Biogeography is the interdisciplinary study of the distribution of species and ecosystems across the planet. It seeks to understand the factors influencing the geographic distribution of organisms, the patterns of species distribution, and the historical processes that have shaped the world's biodiversity. In this comprehensive exploration of biogeography, we will delve into the biogeographical factors influencing ecosystems, patterns of species distribution, and the fascinating realm of historical biogeography.

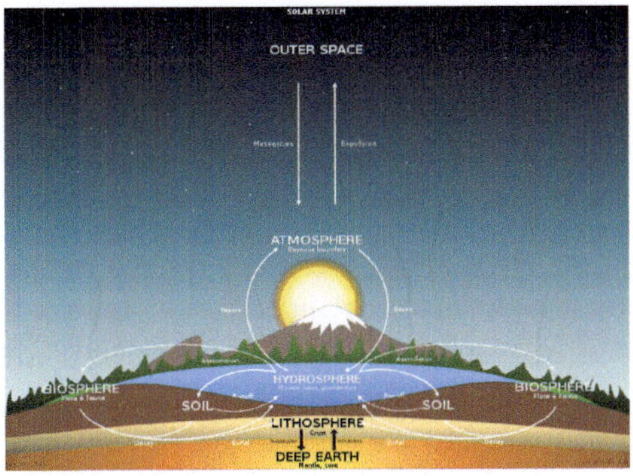

Figure 45. Biogeography

(Source: By Alexander Davronov – Own work, CC BY-SA 4.0, https://commons.wikimedia.org/w/index.php?curid=106124364, Creative Commons License)

Biogeographical Factors Influencing Ecosystems:

- Climate: Climate, including temperature, precipitation, and seasonality, is a primary driver of species distribution. It influences the types of ecosystems that can thrive in a particular region, from tropical rainforests to arctic tundras.
- Geology: Geological features, such as mountains, valleys, and landforms, can create barriers to species movement, leading to distinct biogeographic regions. For example, the Andes Mountains in South America have shaped the distribution of many plant and animal species.
- Topography: The topography of an area, including elevation and slope, affects local climate and habitat diversity. Mountain ranges and valleys, for instance, create microclimates that support unique species assemblages.
- Soil Characteristics: Soil composition and nutrient availability influence the types of vegetation that can grow in an area, which, in turn, affects the species adapted to those habitats.
- Water Bodies: The presence of rivers, lakes, and oceans can act as barriers or corridors for species distribution. Aquatic environments also have their own unique biogeographical patterns.

Patterns of Species Distribution:

- Latitudinal Gradient: One of the most well-known patterns is the latitudinal gradient of species diversity. Generally, biodiversity tends to be highest near the equator and decreases as one moves toward the poles.
- Altitudinal Gradient: Elevation plays a role in species distribution, with different communities of plants and animals occurring at various elevations on mountains. These are often referred to as montane ecosystems.
- Island Biogeography: Islands often have unique species due to their isolation. The theory of island biogeography, developed by Robert MacArthur and E.O. Wilson, explains how species richness on islands is influenced by island size and distance from the mainland.
- Endemism: Some species are found only in specific geographic regions and nowhere else. These are called endemic species and are often associated with isolated habitats.

Historical Biogeography

- Plate Tectonics: The movement of Earth's tectonic plates has had a profound impact on the distribution of species over geological time scales. The separation and collision of continents have created land bridges and barriers, influencing species dispersal.
- Ice Ages: Cycles of glaciation and deglaciation during the Pleistocene epoch shaped the distribution of species by creating and altering habitats. Ice ages allowed species to move across continents via land bridges and also led to the isolation of populations.
- Continental Drift: The concept of continental drift, proposed by Alfred Wegener, explains how continents have shifted over geological time, affecting the distribution of species. For example, the breakup of Pangaea led to the formation of distinct biogeographic regions.
- Evolutionary Radiations: Historical biogeography also explores the diversification of species through adaptive radiations, where a single ancestral species gives rise to many new species, often in response to changes in the environment.

- Biogeography provides valuable insights into the distribution of life on Earth and helps us understand the ecological and evolutionary processes that have shaped the planet's biodiversity. It is a dynamic field that continues to uncover new patterns and connections in the intricate tapestry of life across different regions and time periods.

7.6. ECOLOGICAL DESIGN: HARMONIZING NATURE AND HUMAN ENVIRONMENTS

Ecological design, often referred to as eco-design or eco-friendly design, is a multidisciplinary approach that aims to integrate ecological principles into the planning, design, and construction of human environments. It seeks to create sustainable, harmonious, and resilient spaces that not only meet human needs but also support and enhance the natural world. In this exploration of ecological design, we will delve into how it integrates ecology into urban planning, its application in sustainable landscaping and architecture, and the principles of green infrastructure and ecological design.

Integrating Ecology into Urban Planning:

- Urban Ecology: Ecological design recognizes that cities are ecosystems in their own right, with complex interactions between human and natural elements. Urban planners apply ecological principles to design cities that are more sustainable and resilient, considering factors like green spaces, biodiversity, and waste management.
- Green Spaces: Incorporating parks, green roofs, urban forests, and natural corridors into urban planning helps enhance biodiversity, mitigate heat island effects, and improve air quality. These spaces provide habitats for wildlife and recreational areas for residents.
- Sustainable Transportation: Designing urban transportation systems that prioritize public transit, walking, and cycling reduces the environmental footprint of cities, lowers pollution levels, and promotes healthier lifestyles.
- Urban Agriculture: Integrating urban agriculture into urban planning can contribute to local food production, reduce food miles, and increase access to fresh produce while also enhancing urban greenery.

Sustainable Landscaping and Architecture:

- Native Plants: Sustainable landscaping involves the use of native and drought-tolerant plants that require less water, reducing the need for irrigation and promoting biodiversity.
- Rainwater Harvesting: Designing buildings and landscapes to capture and store rainwater for irrigation and other non-potable uses helps conserve water resources.
- Energy Efficiency: Sustainable architecture incorporates passive design strategies, such as orienting buildings for optimal solar gain, using energy-efficient materials, and employing renewable energy sources like solar panels and wind turbines.
- Green Building Materials: The use of environmentally friendly materials, such as recycled or reclaimed materials, reduces the ecological impact of construction.
- Sustainable Construction Practices: Sustainable architecture minimizes waste and pollution during construction and demolition, ensuring that buildings have a lower environmental footprint.

Green Infrastructure and Ecological Design Principles:

- Bioretention Systems: Ecological design incorporates bioretention systems like rain gardens and swales to manage stormwater runoff naturally, filter pollutants, and recharge groundwater.
- Biodiverse Plantings: Green infrastructure designs prioritize planting a variety of native species to support local ecosystems, enhance soil health, and provide food and habitat for wildlife.
- Permeable Surfaces: Using permeable pavement and surfaces in urban design allows rainwater to infiltrate the ground, reducing flooding and replenishing aquifers.
- Ecosystem Services: Ecological design considers the ecosystem services provided by green infrastructure, including air and water purification, carbon sequestration, and temperature regulation.
- Resilience and Adaptation: In the face of climate change, ecological design principles focus on creating resilient urban environments that can adapt to changing conditions, such as rising temperatures, increased precipitation, and sea-level rise.
- Community Engagement: Sustainable design processes often

involve community engagement to ensure that the needs and values of local residents are considered, fostering a sense of ownership and stewardship.

Ecological design represents a holistic approach to creating human environments that not only minimize negative impacts on the environment but actively contribute to the well-being of ecosystems and communities. By integrating ecological principles into urban planning, landscaping, and architecture, ecological design strives to create sustainable, liveable, and resilient spaces that benefit both people and the planet.

7.7. ECOLOGICAL ECONOMICS

Ecological economics is an interdisciplinary field that seeks to integrate ecological principles and knowledge into economic theory and decision-making. It recognizes that the economy is embedded within the environment and that long-term human well-being is dependent on maintaining the health of ecosystems. In this exploration of ecological economics, we will examine how it links ecology and economics, the concept of valuing ecosystem services, and its role in sustainable resource management.

Linking Ecology and Economics:

- Environmental Economics: Ecological economics extends the principles of environmental economics by emphasizing the interconnectedness of economic systems and ecosystems. It considers the environmental impacts of economic activities and strives to find ways to internalize environmental costs.
- Systems Thinking: Ecological economics incorporates systems thinking, recognizing that economic systems are subsystems of larger ecological systems. It considers feedback loops, thresholds, and complex interactions when analyzing economic and environmental issues.
- Steady-State Economy: Unlike traditional economic growth models, ecological economics advocates for a steady-state economy that maintains a balance between resource use and regeneration, minimizing ecological overshoot.
- Resilience and Adaptation: Ecological economics places a strong emphasis on resilience and adaptive management, recognizing the need to build economic systems that can withstand environmental shocks and changes.

Valuing Ecosystem Services:

- Ecosystem Services Framework: Ecological economics employs the concept of ecosystem services, which assigns value to the benefits that ecosystems provide to human well-being. These services include provisioning (e.g., food, water), regulating (e.g., climate regulation), supporting (e.g., nutrient cycling), and cultural (e.g., recreational) services.
- Natural Capital: Ecological economics views ecosystems as natural capital, highlighting their importance as a source of wealth and well-being. This perspective emphasizes the need to manage natural capital sustainably to ensure long-term economic and ecological health.
- Externalities: Ecological economics addresses environmental externalities by considering the ecological and social costs of economic activities. It seeks to internalize these costs through policy mechanisms like carbon pricing and pollution taxes.
- Discounting: Ecological economics challenges the use of high discount rates in traditional economics, which can lead to undervaluing long-term ecological and social benefits. It promotes alternative approaches like sustainability discounting.

Sustainable Resource Management:

- Resource Depletion: Ecological economics recognizes the finite nature of natural resources and advocates for responsible and sustainable resource management. It calls for the preservation of critical ecosystem services, such as clean water, fertile soil, and biodiversity.
- Common-Pool Resources: Ecological economics explores the management of common-pool resources, such as fisheries and forests, using concepts like the tragedy of the commons and adaptive co-management to ensure sustainable use.
- Degrowth: In response to environmental challenges, some ecological economists advocate for degrowth—a deliberate reduction in economic production and consumption to alleviate ecological strain and enhance well-being.
- Alternative Indicators: Ecological economics promotes the use of alternative indicators of well-being, such as the Genuine Progress

Indicator (GPI) and the Ecological Footprint, which provide a more comprehensive assessment of societal progress beyond traditional economic measures like Gross Domestic Product (GDP).

- Environmental Policy: Ecological economics contributes to the development of environmental policies that aim to balance economic development with ecological sustainability. Policies may include cap-and-trade systems, emissions reduction targets, and conservation programs.

Ecological economics serves as a bridge between ecology and economics, offering a framework for reevaluating economic goals and practices in light of ecological constraints. By valuing ecosystem services, promoting sustainable resource management, and integrating ecological principles into economic decision-making, ecological economics strives to create a more sustainable and equitable future for both humanity and the planet.

7.8. ECOLOGICAL ENGINEERING: NATURE-INSPIRED SOLUTIONS FOR A SUSTAINABLE FUTURE

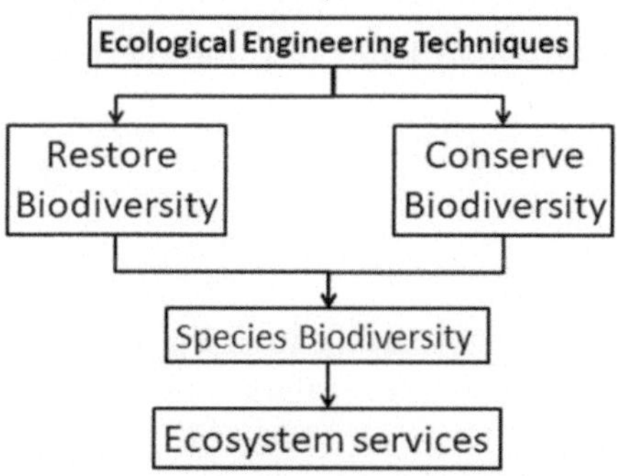

Figure 46. Ecological Engineering

(Source: Heong, K.-L., Lu, Z.-X., Chien, H.-V., Escalada, M., Settele, J., Zhu, Z.-R., Cheng, J.-A. Ecological Engineering for Rice Insect Pest Management: The Need to Communicate Widely, Improve Farmers' Ecological Literacy and Policy Reforms to Sustain Adoption. Agronomy. 2021; 11(11), 2208. https://doi.org/10.3390/agronomy11112208, Creative Commons License)

Ecological engineering is an innovative field that draws inspiration from natural ecosystems to design and implement sustainable solutions for various environmental challenges. It seeks to harmonize human activities with the environment, addressing issues such as pollution, habitat degradation, and ecosystem restoration. In this exploration of ecological engineering, we will examine how it harnesses nature-inspired solutions, its applications in environmental remediation, and the challenges and ethical considerations associated with its practice.

Engineering Solutions Inspired by Nature:

Biomimicry: Ecological engineering applies principles of biomimicry, where engineers and scientists observe and replicate natural processes and structures to create sustainable solutions. Examples include designing buildings that mimic termite mounds' temperature regulation or creating self-cleaning surfaces inspired by lotus leaves.

Ecosystem Services: Ecological engineering recognizes the value of ecosystem services, such as water purification, pollination, and flood control, and seeks to design human systems that emulate or enhance these services.

Restoration Ecology: The field incorporates concepts from restoration ecology, aiming to restore or rehabilitate damaged ecosystems by designing interventions that accelerate recovery processes and support native species.

Green Infrastructure: Ecological engineering promotes the use of green infrastructure, including constructed wetlands, vegetated swales, and green roofs, to manage stormwater, reduce urban heat island effects, and enhance urban biodiversity.

Applications in Environmental Remediation:

Wastewater Treatment: Ecological engineering designs natural or constructed wetlands to treat wastewater, using plants and microorganisms to remove pollutants and nutrients. This approach provides efficient and cost-effective treatment while creating a habitat for wildlife.

- Soil Remediation: Phytoremediation, a technique within ecological engineering, uses plants to remove contaminants from soil. Certain plant species can accumulate heavy metals or break down organic pollutants.
- Erosion Control: Ecological engineering applies bioengineering techniques like planting native vegetation along shorelines or riverbanks to prevent erosion, stabilize soil, and enhance habitat.
- Habitat Creation: The field designs and constructs habitats, such as oyster reefs, artificial islands, or fish ladders, to support

biodiversity and restore ecosystems impacted by human activities.

- Eco-Friendly Agriculture: Ecological engineering encourages practices like agroforestry, which combines tree planting with crop cultivation to improve soil fertility, reduce erosion, and enhance biodiversity.

Challenges and Ethical Considerations:

- Trade-Offs: Balancing human needs and ecological goals can be challenging. Ecological engineering often involves trade-offs between economic development and environmental conservation.
- Unintended Consequences: Implementing nature-inspired solutions may have unintended consequences. Introducing non-native species for restoration purposes, for example, can lead to invasive species issues.
- Ethical Considerations: Ethical questions arise regarding human intervention in natural systems. Ecological engineering prompts discussions about the boundaries of acceptable manipulation of ecosystems and the potential for unintended harm.
- Scale and Complexity: Some environmental challenges are vast and complex, making it difficult to design effective ecological engineering solutions at the necessary scale.
- Long-Term Monitoring: Sustaining the benefits of ecological engineering interventions often requires long-term monitoring and management, which can be resource-intensive.
- Regulatory and Policy Frameworks: Developing appropriate regulations and policies to govern ecological engineering projects can be challenging. Ensuring that interventions comply with environmental laws and ethical standards is essential.

Ecological engineering represents a promising approach to address pressing environmental issues by drawing inspiration from nature's solutions. While it offers innovative and sustainable solutions to complex problems, it also raises important ethical and practical considerations that require thoughtful and responsible implementation. As the field continues to evolve, it holds the potential to play a significant role in fostering a more harmonious relationship between humanity and the natural world.

7.9. ECOLOGICAL ANTHROPOLOGY: BRIDGING CULTURES AND ENVIRONMENTS

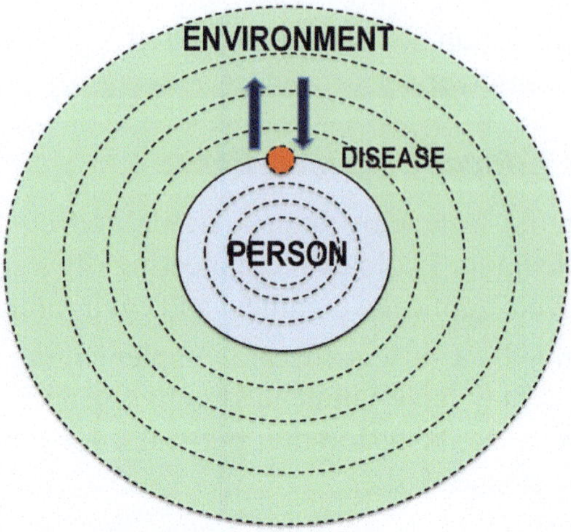

Figure 47. Ecological Anthropology

(Source: Tretter, F, Löffler-Stastka, H. The Human Ecological Perspective and Biopsychosocial Medicine. International Journal of Environmental Research and Public Health. 2019; 16(21), 4230. https://doi.org/10.3390/ijerph16214230 , Creative Commons License)

Ecological anthropology is a subfield of anthropology that explores the intricate relationship between human societies and their environments, focusing on how cultural practices, beliefs, and behaviors influence and are influenced by the natural world. It sheds light on the complex interplay between human cultures and ecosystems, offering insights into sustainable practices, environmental impacts, and the connection between cultural celebrations and nature. In this exploration of ecological anthropology, we will delve into the intersection of anthropology and ecology, cultural practices and environmental impact, and the concept of festive ecology.

Intersection of Anthropology and Ecology

Holistic Perspective: Ecological anthropology adopts a holistic perspective, acknowledging that human societies are embedded within their ecosystems. It considers the social, economic, cultural, and ecological dimensions of human-environment interactions.

- Cultural Ecology: Cultural ecology, a key component of

ecological anthropology, examines how cultural beliefs, practices, and knowledge systems shape human adaptation to local environments. It explores the ways in which societies manage resources, engage in subsistence strategies, and adapt to environmental change.

- Ethnographic Research: Ecological anthropologists conduct ethnographic research to study the lifeways of different cultures, documenting their relationships with the natural world. This research often involves long-term immersion in communities to gain a deep understanding of their ecological practices.

Cultural Practices and Environmental Impact

- Subsistence Strategies: Ecological anthropology investigates how various cultures obtain food, water, and other resources from their environments. It explores practices like hunting and gathering, agriculture, pastoralism, and fishing, and assesses their environmental consequences.
- Indigenous Knowledge: The field recognizes the value of indigenous and traditional ecological knowledge, which often includes sustainable resource management practices that have evolved over generations.
- Environmental Change: Ecological anthropology explores how human activities, such as deforestation, overfishing, or urbanization, can lead to environmental degradation and impact the livelihoods of local communities.
- Conservation and Development: It engages with questions of conservation and development, examining the tensions between preserving ecosystems and supporting human well-being, especially in regions rich in biodiversity.

Festive Ecology: Understanding Cultural Celebrations and Nature

Festive Traditions: Festive ecology explores how cultural celebrations and rituals are intertwined with nature. Many festivals are linked to agricultural cycles, seasonal changes, and natural events like harvests or migrations.

- Cultural Identity: Festive ecology examines how cultural traditions and festivals shape collective identity, foster community bonds, and strengthen connections to the land and environment.

- Sustainable Practices: It highlights the role of festive practices in promoting sustainability. Some celebrations involve rituals that reinforce responsible resource use, such as communal fishing ceremonies that limit catch size.
- Ecological Knowledge Transmission: Festive ecology considers the transmission of ecological knowledge and values from one generation to the next through storytelling, rituals, and celebrations.

Ecological anthropology offers valuable insights into the ways in which human cultures interact with and impact the natural world. By studying cultural practices, environmental knowledge systems, and the relationships between humans and their environments, ecological anthropologists contribute to our understanding of sustainability, resilience, and the complex dynamics of human societies in an ever-changing world.

7.9.1. Festive Ecology: Celebrating Nature and Culture

Festive ecology is a fascinating aspect of ecological anthropology that explores the intricate relationship between cultural celebrations, rituals, and the natural environment. It delves into how human societies have historically marked significant events, seasons, and transitions by drawing inspiration from the rhythms of nature. In this exploration of festive ecology, we will uncover the profound connections between cultural festivities and the natural world, shedding light on the ways in which societies have celebrated and revered nature throughout history.

Celebration of Seasonal Changes:

Harvest Festivals: Many cultures around the world celebrate the bountiful harvests of crops as a way to give thanks for the earth's abundance. These festivals often coincide with the changing seasons and are marked by feasts, music, and dance.

Equinox and Solstice Celebrations: Equinoxes and solstices, representing significant astronomical events, have been celebrated by various cultures. They mark the changing lengths of days and nights and are often tied to planting and harvest times.

Spring Festivals: Spring festivals celebrate the renewal of life in nature, with traditions like egg hunting, flower parades, and maypole dancing symbolizing the reawakening of the natural world after winter.

Cultural Rituals and Nature:

Religious Ceremonies: Many religious traditions incorporate nature into their rituals. For example, Hinduism's Diwali, the Festival of Lights, often involves the lighting of oil lamps to symbolize the victory of light over darkness and the return of knowledge and righteousness.

Animistic Beliefs: Indigenous cultures often practice animism, where they believe that all elements of nature possess spirits or souls. Festive rituals in animistic cultures frequently involve offerings to appease or celebrate the spirits of nature.

Nature-Based Spirituality: Contemporary nature-based spiritual movements, like Wicca or modern Paganism, emphasize a deep connection to nature and celebrate natural cycles through festivals like the Wheel of the Year.

Conservation and Sustainability:

Environmental Awareness: Festive ecology can promote environmental awareness and conservation. Some celebrations have evolved to include eco-friendly practices, such as using biodegradable materials for decorations or organizing clean-up efforts in natural areas.

Cultural Preservation: Festivals that celebrate nature often play a role in preserving cultural traditions and ecological knowledge, ensuring that they are passed down to future generations.

Community Engagement: Festive ecology fosters community engagement and a sense of belonging by bringing people together to celebrate shared cultural and ecological values.

Ethical Considerations: Some modern celebrations have raised ethical questions about their impact on the environment, particularly when they involve activities that generate waste or disrupt local ecosystems.

Festive ecology underscores the profound connections between human cultures and the natural world. It reveals how societies have historically marked the passage of time, the changing of seasons, and significant life events by drawing inspiration from the environment. In a world facing environmental challenges, festive ecology offers a lens through which to explore sustainable and meaningful ways of celebrating and reconnecting with nature.

7.10. ECOLOGICAL HEALTH: A HOLISTIC APPROACH TO HUMAN AND ECOSYSTEM WELL-BEING

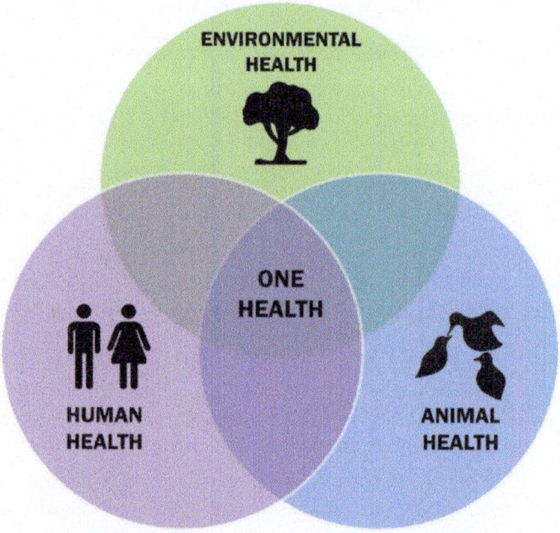

Figure 48. Ecological Health and Ecosystem Well-being

(Source: By Thddbfk – Own work, CC BY-SA 4.0, https://commons.wikimedia.org/w/index.php?curid=81872126, Creative Commons License)

Ecological health is a multidisciplinary field that recognizes the intricate interconnections between human health and the well-being of ecosystems. It emphasizes the importance of maintaining ecological balance for the benefit of both the environment and human populations. In this exploration of ecological health, we will examine the relationships between human health and ecosystem well-being, ecohealth approaches to disease prevention, and the use of ecological indicators to assess health.

Human Health and Ecosystem Well-being:

- Ecosystem Services: Ecological health recognizes that ecosystems provide essential services that directly or indirectly influence human health, including clean air and water, food, medicinal plants, and natural spaces for physical and mental well-being.
- Biodiversity: Biodiversity is crucial for human health, as diverse ecosystems can help regulate diseases, provide nutritional variety, and offer potential sources of new medicines.
- Resilience: Healthy ecosystems are more resilient to disturbances, such as climate change or disease outbreaks, which can indirectly benefit human communities by reducing vulnerability.

- Access to Nature: Access to natural environments, such as parks and green spaces, has been linked to improved mental health, reduced stress, and increased physical activity.

Ecohealth Approaches to Disease Prevention:

- One Health: The One Health approach recognizes that the health of humans, animals, and ecosystems are interconnected. It seeks to address zoonotic diseases (those transmitted between animals and humans) by considering the environmental factors that facilitate disease transmission.
- Ecosystem Restoration: Restoring degraded ecosystems can reduce disease risk. For example, restoring wetlands can help control mosquito populations and reduce the spread of diseases like malaria.
- Agroecology: Sustainable agricultural practices that promote ecological balance, such as organic farming and crop rotation, can reduce the need for chemical pesticides and herbicides, which can harm human health.
- Urban Planning: Ecohealth approaches to urban planning emphasize designing cities with green infrastructure and sustainable transportation options that support both human health and the environment.

Ecological Indicators of Health:

- Biomonitoring: Biomonitoring involves the use of biological indicators, such as certain species or populations, to assess the health of ecosystems. For example, the presence of specific aquatic organisms can indicate water quality.
- Ecosystem Services Assessment: Ecological health assessments consider the provision of ecosystem services and the impacts of ecosystem degradation on human well-being. Economic valuations may be used to demonstrate the benefits of healthy ecosystems.
- Environmental Quality Metrics: Measures of environmental quality, such as air and water quality indices, can provide insights into the health of ecosystems and their potential impacts on human health.

- Biodiversity Metrics: Assessments of biodiversity, including species richness and genetic diversity, can serve as indicators of ecosystem health and resilience.
- Disease Surveillance: Monitoring the incidence and prevalence of diseases, especially zoonotic diseases, can provide insights into the health of ecosystems and the potential risks to human populations.
- Ecological health represents a holistic and forward-thinking approach that recognizes the inextricable links between human health and the environment. By promoting the conservation and restoration of ecosystems, adopting ecohealth approaches to disease prevention, and using ecological indicators to assess health, we can work towards a future where both ecosystems and human communities thrive in balance and harmony.

7.11. ECOSOPHY/ECOPHILOSOPHY

Ecosophy, also known as ecophilosophy, is a branch of philosophy that delves into the profound connections between human beings, the natural world, and the ethical and philosophical implications of ecological thinking. It provides a philosophical foundation for understanding humanity's place in the ecosystem and advocates for a more harmonious and sustainable relationship with the environment. In this exploration of ecosophy/ecophilosophy, we will examine its philosophical foundations, the concepts of deep ecology and ecological ethics, and the personal and societal implications of this philosophical perspective.

Philosophical Foundations of Ecological Thinking

- Holism: Ecosophy emphasizes a holistic worldview, viewing the natural world as interconnected and interdependent. It challenges reductionist approaches that isolate components of ecosystems and highlights the need for a comprehensive understanding of ecological systems.
- Interconnectedness: Ecosophy explores the idea that all living and non-living entities are interconnected in a complex web of relationships. It considers the ecological, social, and ethical consequences of these interconnections.
- Systems Thinking: Ecosophy draws upon systems thinking,

which views ecological systems as dynamic and self-regulating. It recognizes that changes in one part of an ecosystem can have far-reaching consequences throughout the system.

Deep Ecology and Ecological Ethics

Deep Ecology: Deep ecology is a philosophical and environmental movement associated with ecosophy/ecophilosophy. It advocates for a fundamental shift in human consciousness, moving from anthropocentrism (human-centered) to ecocentrism (ecosystem-centered). It posits that all life has intrinsic value, not merely instrumental value to humans.

Biocentrism: Ecosophy often promotes a biocentric ethic, which extends ethical consideration to all living beings, not just humans. It challenges the traditional anthropocentric view that places human interests above all others.

Intrinsic Value: Ecosophy raises questions about the intrinsic value of nature, suggesting that nature has inherent worth apart from its utility to humans. This perspective challenges the commodification of nature.

Personal and Societal Implications

- Ethical Action: Ecosophy encourages individuals and societies to align their actions with ecological ethics. This includes making choices that prioritize environmental sustainability, conservation, and respect for all life forms.
- Environmental Advocacy: Ecosophy often leads to advocacy for environmental protection and conservation efforts. It motivates individuals and organizations to work towards policies and practices that safeguard ecosystems.
- Cultural and Spiritual Transformation: Ecosophy has influenced cultural and spiritual movements that emphasize a deeper connection to nature. Practices like eco-spirituality and eco-mindfulness encourage a more profound appreciation of the natural world.
- Political and Policy Change: Ecosophy has contributed to the development of ecocentric policy frameworks, such as the Rights of Nature movement, which seeks to legally recognize the rights of ecosystems and natural entities.
- Sustainable Living: Ecosophy encourages sustainable living practices, including reduced consumption, responsible resource

use, and support for regenerative agriculture and renewable energy sources.

- Environmental Education: Ecosophy advocates for environmental education that incorporates philosophical and ethical considerations, helping individuals understand the broader implications of their choices.

Ecosophy/ecophilosophy offers a philosophical framework that challenges conventional anthropocentric perspectives and calls for a profound shift in human consciousness towards ecological thinking. It prompts individuals and societies to reevaluate their relationship with the natural world, embrace ecological ethics, and take actions that promote a more sustainable and harmonious coexistence with the environment.

7.12. ENVIRONMENTAL PSYCHOLOGY

Environmental psychology is an interdisciplinary field that explores the dynamic interplay between individuals and their physical, social, and natural environments. It seeks to understand how human behavior, attitudes, and well-being are influenced by the environment and how environmental design can shape human experiences and choices. In this section of environmental psychology, we will delve into the understanding of human behavior and its environmental impact, the concept of ecotherapy and its effects on human well-being, and the principles of designing eco-friendly environments.

Understanding Human Behaviour and Environmental Impact:

- Proximate and Ultimate Factors: Environmental psychology examines both proximate and ultimate factors that influence human behavior. Proximate factors include immediate situational cues, while ultimate factors consider long-term evolutionary, cultural, and psychological influences.
- Environmental Attitudes: The field investigates how individuals' attitudes towards the environment, including beliefs about nature's value and our role in it, shape behaviors like conservation efforts, resource consumption, and recycling.
- Environmental Stressors: Environmental psychology studies the psychological and physiological effects of environmental stressors, such as noise, pollution, and overcrowding, on human well-being and behavior.

- Place Attachment: It explores the emotional and psychological bonds people form with places, including their homes, neighborhoods, and natural landscapes, and how these attachments influence behaviors and decisions.

Ecotherapy and Human Well-being:

- Nature as a Healing Agent: Ecotherapy, also known as nature therapy or green therapy, examines the therapeutic benefits of nature exposure. It suggests that spending time in natural environments can reduce stress, anxiety, and depression, while enhancing overall mental and emotional well-being.
- Restorative Environments: Environmental psychology identifies restorative environments—settings that promote psychological restoration, attention recovery, and stress reduction. Natural settings, such as parks and forests, are often considered restorative due to their calming and rejuvenating effects.
- Biophilic Design: Biophilic design principles integrate natural elements and patterns into built environments, promoting well-being and cognitive performance. This approach recognizes the innate human connection to nature and incorporates it into architectural and interior design.
- Therapeutic Landscapes: Environmental psychologists study therapeutic landscapes, which are environments designed to support healing and well-being. These may include hospital gardens, meditation spaces, or outdoor rehabilitation areas.

Designing Eco-friendly Environments:

- Sustainable Architecture: Environmental psychology informs sustainable architecture and urban planning by considering the impact of built environments on human behavior and health. It emphasizes energy-efficient buildings, green infrastructure, and pedestrian-friendly designs.
- Behavioral Interventions: The field explores how behavioral interventions, such as feedback mechanisms and social norms, can encourage eco-friendly behaviors like energy conservation, waste reduction, and public transportation use.
- Environmental Communication: Environmental psychology

contributes to effective environmental communication strategies by addressing cognitive biases, risk perception, and the role of emotions in decision-making related to environmental issues.

- User-Centered Design: It emphasizes user-centered design approaches that involve the input and preferences of individuals to create spaces that meet their needs and foster sustainable behaviors.
- Community Engagement: Environmental psychologists work with communities to facilitate engagement in environmental decision-making processes, such as urban planning initiatives and conservation efforts.

Environmental psychology offers valuable insights into how the environment influences human behavior and well-being, and how design and interventions can promote eco-friendly behaviours and healthier living spaces. By understanding these dynamics and applying them to various contexts, the field contributes to creating environments that benefit both individuals and the planet.

7.13. NATURAL HISTORY

Natural history, a discipline rooted in the observation and documentation of the natural world, plays a foundational role in ecology. It serves as the basis for ecological research, enabling scientists to understand the intricate relationships between organisms and their environments. In this exploration of natural history, we will examine its critical role in ecology, the importance of field observation and documentation, and its significant contributions to ecological knowledge.

Role of Natural History in Ecology

Historical Context: Natural history has deep historical roots, with naturalists like Charles Darwin, Alfred Russel Wallace, and Carl Linnaeus contributing foundational observations and concepts to the development of ecological science.

- Observation and Description: Natural history is rooted in the meticulous observation and description of the natural world. It involves documenting the behaviors, life histories, and interactions of organisms, as well as the characteristics of their environments.

- Holistic Understanding: Natural history fosters a holistic understanding of ecosystems. It considers the interconnections between species, the flow of energy and nutrients, and the physical and biological factors that shape ecosystems.
- Context for Ecological Research: Ecological research often begins with natural history observations. These observations provide a context for generating hypotheses and designing experiments.

Field Observation and Documentation

- Direct Observation: Naturalists and ecologists conduct direct field observations to gather data on species' behaviors, feeding habits, reproduction, and responses to environmental changes.
- Specimen Collection: Specimen collection, including plant samples, insect specimens, and soil samples, is a crucial part of natural history. Collections serve as references for species identification and contribute to the understanding of biodiversity.
- Photography and Recording: Modern technology, such as digital photography and audio recordings, has enhanced the ability to document natural history observations, enabling detailed records and visual evidence.
- Citizen Science: Citizen science projects engage the public in natural history data collection, contributing valuable data for ecological research. Examples include birdwatching programs and plant identification initiatives.

Contributions to Ecological Knowledge

- Species Discovery: Natural history has led to the discovery of numerous species, many of which are essential components of ecosystems. Understanding species diversity is fundamental to ecology.
- Life History Patterns: Natural history observations provide insights into the life history patterns of organisms, including reproductive strategies, migration patterns, and seasonal behaviors.
- Ecological Interactions: Naturalists have documented intricate ecological interactions, such as predator-prey relationships, mutualisms, and competition, which form the foundation of ecological theory.

- Conservation and Biodiversity: Natural history data contribute to conservation efforts by identifying species at risk, understanding their habitat requirements, and assessing the health of ecosystems.
- Long-Term Monitoring: Natural history records provide valuable long-term data that can reveal trends and changes in ecosystems over time, helping scientists understand the impacts of environmental change.
- Educational Value: Natural history collections and observations are vital for educating the public and future generations about the natural world, fostering a sense of wonder and appreciation for biodiversity.

Natural history is not merely a historical discipline but an ongoing and evolving field that continues to inform and enrich ecological research. Its emphasis on direct observation, documentation, and a deep understanding of the natural world remains integral to advancing our knowledge of ecosystems and their complex dynamics.

7.14. SYSTEMS ECOLOGY

Systems ecology is a branch of ecology that takes a holistic and interdisciplinary approach to studying ecosystems as complex, interconnected systems. It emphasizes understanding the interactions and feedbacks among components of ecosystems, including living organisms, physical processes, and biogeochemical cycles. In this exploration of systems ecology, we will delve into its holistic approaches to studying ecosystems, the use of ecosystem modeling and simulation, and the role of complex adaptive systems in ecology.

Holistic Approaches to Studying Ecosystems:

Ecosystem as a Whole: Systems ecology views an ecosystem as a whole entity rather than a collection of individual components. It considers the interactions between organisms, the flow of energy and matter, and the dynamic processes that shape the ecosystem.

- Interconnectedness: The field recognizes the interconnectedness of ecological components. It explores how changes in one part of an ecosystem can have cascading effects throughout the system, affecting both biotic and abiotic factors.
- Emergent Properties: Systems ecology seeks to identify emergent

properties—properties that arise from the interactions of components and are not evident when studying components in isolation. These emergent properties can include self-regulation and resilience.

Ecosystem Modelling and Simulation:

- Mathematical Models: Systems ecologists use mathematical models to represent ecological processes and interactions. These models can range from simple equations to complex simulations that incorporate multiple variables and feedback loops.
- Predictive Tools: Ecosystem models serve as predictive tools, allowing scientists to simulate how ecosystems might respond to changes in environmental conditions, human interventions, or disturbances.
- Scenario Analysis: By running scenarios in ecosystem models, researchers can explore potential outcomes and make informed decisions about ecosystem management, conservation, and restoration.
- Feedback Loops: Ecosystem models often incorporate feedback loops, which represent how changes in one component of the ecosystem can influence other components and, in turn, feedback to the original component.

Complex Adaptive Systems in Ecology:

- Adaptation and Evolution: Systems ecology draws on principles of complex adaptive systems theory to understand how ecosystems adapt and evolve over time. It considers how species and ecosystems respond to environmental changes and selective pressures.
- Self-Organization: Complex adaptive systems theory highlights the self-organizing properties of ecosystems. Ecosystems can exhibit emergent patterns and structures through self-organization, such as the formation of spatial patterns in vegetation.
- Resilience: Resilience is a key concept in complex adaptive systems. Systems ecology explores how ecosystems can absorb disturbances, adapt to changing conditions, and maintain their functions and structures.

- Nonlinear Dynamics: Ecosystems often display nonlinear dynamics, where small changes in one variable can lead to disproportionately large effects. These dynamics can result in ecological surprises and regime shifts.
- Cross-Scale Interactions: Complex adaptive systems theory considers interactions across different spatial and temporal scales. It recognizes that processes at one scale can influence processes at other scales, leading to scale-dependent behaviors.

Systems ecology provides valuable insights into the dynamics and behavior of ecosystems, offering a holistic perspective that considers the complex interactions and feedback that shape ecological systems. By using mathematical models, simulations, and principles of complex adaptive systems, systems ecologists contribute to our understanding of how ecosystems function, respond to change, and can be managed for sustainability and resilience.

7.15. OTHER NON-BIOLOGICAL DISCIPLINES IN ECOLOGY: EXPLORING DIVERSE ECOLOGICAL PERSPECTIVES

While biology plays a central role in ecological studies, various non-biological disciplines contribute unique perspectives to the field of ecology. These disciplines emphasize different aspects of ecological interactions, extending beyond the realm of living organisms. In this exploration of non-biological disciplines in ecology, we will delve into media ecology, industrial ecology, and information ecology, highlighting their distinct focuses and contributions to ecological understanding.

Media Ecology

- Communication Ecosystems: Media ecology extends ecological thinking to communication systems and technologies, considering how media and communication channels function as ecosystems. It explores the flow of information, ideas, and messages within these ecosystems.
- Effects on Human Behaviour: Media ecology examines how different forms of media, such as television, the internet, and social media, influence human behavior, values, and perception of the natural world. It considers the ecological impacts of media

consumption on societal attitudes and behaviors towards the environment.

- Media as an Environment: In the context of media ecology, the media environment is viewed as a complex ecosystem where various media platforms, content, and technologies interact. The ecological approach helps analyze the relationships and feedback loops within this environment.

Industrial Ecology

- Mimicking Natural Systems: Industrial ecology applies ecological principles to industrial systems, emphasizing the concept of "industrial ecosystems." It seeks to create industrial processes and systems that mimic natural ecosystems in terms of efficiency, resource use, and waste management.
- Resource Flows: Industrial ecology analyzes the flows of materials, energy, and waste within industrial systems, aiming to optimize resource use, minimize waste generation, and reduce environmental impacts.
- Circular Economy: The field promotes the concept of a circular economy, where materials are reused, recycled, and repurposed rather than discarded as waste. It aligns with the ecological principles of sustainability and minimizing environmental degradation.

Information Ecology

- Information Ecosystems: Information ecology explores the ecosystems of information, knowledge, and data flows within human societies. It considers how information is generated, transmitted, and processed, akin to the flow of energy and nutrients in natural ecosystems.
- Information Pollution: Just as ecosystems can suffer from pollution, information ecosystems can be affected by "information pollution." This includes misinformation, disinformation, and the overabundance of low-quality information, which can have ecological analogs in terms of ecosystem degradation.
- Knowledge Management: Information ecology is concerned with the management and stewardship of knowledge resources.

It examines how societies organize and disseminate information to support decision-making and problem-solving, similar to how ecosystems manage resources.

These non-biological disciplines in ecology broaden our understanding of ecological systems beyond traditional biological perspectives. They highlight the interconnectedness of human activities, information flows, and industrial processes with the natural environment. By applying ecological thinking to these diverse disciplines, we gain insights into the complex relationships between society, technology, communication, and the ecosystems we inhabit.

7.16. REFERENCES

1. Alexander, K. A., Hobday, A. J., Cvitanovic, C., Ogier, E., Nash, K., Cottrell, R. S., Fleming, A., Fudge, M., Fulton, E. A., Frusher, S., Kelly, R., Macleod, C., Pecl, G. T., van Putten, E. I., Vince, J., Watson, R. A., (2018). Progress in integrating natural and social science in marine ecosystem-based management research. Mar Freshw Res 70, 71–83.
2. Bammer, G., (2016). What constitutes appropriate peer review for interdisciplinary research? Palgrave Commun 2, 16017.
3. Blythe, J., Nash, K., Yates, J., Cumming, G. S., (2017). Feedbacks as a bridging concept for advancing transdisciplinary sustainability research. Curr Opin Environ Sustain 26–27,114–119.
4. Bracken, L. J., Oughton, E. A., (2006). 'What do you mean?' The importance of languge in developing interdisciplinary research. Trans Inst Br Geogr 31, 371–382.
5. Bridle, H., Vrieling, A., Cardillo, M., Araya, Y., Hinojosa, L., (2013). Preparing for an interdisciplinary future: a perspective from early-career researchers. Futures 53, 22–32.
6. Bromham, L., Dinnage, R., Hua, X., (2016). Interdisciplinary research has consistently lower funding success. Nature 534, 684–687.
7. Budtz Pedersen, D., (2016). Integrating social sciences and humanities in interdisciplinary research. Palgrave Commun 2, 16036.
8. Cabana, M. D., Rand, C. S., Powe, N. R., Wu, A. W., Wilson, M. H., Abboud, P. A. C., Rubin, H. R., (1999). Why don't physicians follow clinical practice guidelines? A framework for improvement. JAMA 282, 1458–1465.
9. Campbell, L. M., (2005). Overcoming obstacles to interdisciplinary research. Conserv Biol 19, 574–577.
10. Chapman, J. M., Algera, D., Dick, M., Hawkins, E. E., Lawrence, M. J., Lennox, R. J., Rous, A. M., Souilere, C. M., Stemberger, H. L. J., Struthers, D. P., Vu, M., Ward, T. D., Zolderdo, A. J., Cooke, S. J., (2015). Being relevant: practical guidance to early career researchers interested in solving conservation problems. Glob Ecol Conserv 4, 334–348.
11. Charnley, S., Carothers, C., Satterfield, T., Levine, A., Poe, M. R., Norman, K., Donatuto, J., Breslow, S. J., Mascia, M. B., Levin, P. S., Basurto, X., Hicks, C. C., García-Quijano, C., St. Martin, K., (2017). Evaluating the best available social science for natural resource

management decision-making. Environ Sci Policy 73, 80–88.

12. Clark, S. G., Steelman, T. A., (2013). Interviewing for an interdisciplinary job: principled goals, pragmatic outcomes, and finding the right fit in academia. J Environ Stud Sci 3, 21–29.
13. Coen, D. R., (2018). Climate in motion: science, empire and the problem of scale. The University of Chicago Press, Chicago.
14. Cvitanovic, C., McDonald, J., Hobday, A. J., (2016). From science to action: principles for undertaking environmental research that enables knowledge exchange and evidence-based decision-making. J Environ Manag 183, 864–874.
15. Dick, M., Rous, A. M., Nguyen, V. M., Cooke, S. J., (2016). Necessary but challenging: multiple disciplinary approaches to solving conservation problems. FACETS 1, 62–82.
16. Eigenbrode, S. D., O'Rourke, M., Wulfhorst, J. D., Althoff, D. M., Goldberg, C. S., Merrill, K., Morse, W., Nielsen-Pincus, M., Stephens, J., Winowiecki, L., Bosque-Pérez, N. A., (2007). Employing philosophical dialogue in collaborative science. Bioscience 57, 55–64.
17. Evans, M. C., Cvitanovic, C., (2018). An introduction to achieving policy impact for early career researchers. Palgrave Commun 4, 88. https://doi, Biggs, R., Norström, A. V., Reyers, B., Rockstrom, J., (2016). Socio-ecological resilience and biosphere-based sustainability science. Ecol Soc 21(3), 41. https://doi, Bergstrom, C. T., Börner, K., Evans, J. A., Helbing, D., Milojevič, S., Petersen, A. M., Radicchi, F., Sinatra, R., Uzzi, B., Vespignani, A., Waltman, L., Wang, D., Barabási A-L (2018). Science of science. Science. https://doi, Hobday, A. J., Jennings, S., Creighton, C., D'Silva, D., Haward, M., Holbrook, N. J., Nursey-Bray, M., Pecl, G. T., van Putten, E. I., (2014). The short history of research in a marine climate change hotspot: from anecdote to adaptation in south-east Australia. Rev Fish Biol Fisheries 24, 596 611.
18. Glaser, B., Strauss, A., (1967). The discovery of grounded theory: strategies for qualitative research. Aldine, Chicago.
19. Haider, L. J., Hentati-Sundberg, J., Giusti, M., Goodness, J., Hamann, M., Masterson, V. A., Meacham, M., Merrie, A., Ospina, D., Schill, C., Sinare, H., (2018). The undisciplinary journey: early-career perspectives in sustainability science. Sustain Sci 13, 191–204.
20. Hall, K. L., Stokols, D., Stipelman, B. A., Vogel, A. L., Feng, A., Masimore, B., Morgan, G., Moser, R. P., Marcus, S. E., Berrigan, D.,

(2012). Assessing the value of team science: a study comparing center- and investigaro- initiated grants. Am J Prev Med 42, 157–163.

21. Hay, I., (2010). Qualitative research methods in human geography. Oxford University Press, North York.
22. Hein, C. J., Ten Hoeve, J. E., Gopalakrishnan, S., Livneh, B., Adams, H. D., Marino, E., Weiler, C. S., (2018). Overcoming early career barriers to interdisciplinary climate change research. WIREs Clim Change 9, e530.
23. Klein, J. T., (2005). The discourse on transdisciplinarity: an expanding global field. In: Klein, K. T., Grossbacker-Mansuy, W., Haberli, R., Bill, A., Scholz, R. W., Welti, W., (eds) Transdisciplinarity: joint problem solving among science, technology and society: an effective way for managing complexity. Birkhauser, Basel, Switzerland, pp 35–44.
24. Lyall, C., Meagher, L. R., (2012). A masterclass in interdisciplinarity: research into practice in training the next generation of interdisciplinary researchers. Futures 44, 608–617.
25. Marshall, N., Adger, N., Attwood, S., Brown, K., Crissman, C., Cvitanovic, C., De Young, C., Gooch, M., James, C., Jessen, S., Johnson, D., Marshall, P., Park, S., Wachenfeld, D., Wrigley, D., (2017). Empirically derived guidance for social scientists to influence environmental policy. PLoS ONE 12, e0171950.
26. Meyer, S. R., Levesque, V. R., Hutchins Bieluch, K., Johnson, M. L., McGreavy, B., Dreyer, S., Smith, H., (2016). Sustainability science graduate students as boundary spanners. J Environ Stud Sci 6, 344–353.
27. Morse, W., Nielsen-Pincus, M., Force, J., Wulfhorst, J. D., (2007). Bridges and barriers to developing and conducting interdisciplinary graduate-student team research. Ecol Soc 12, 1–14.
28. Nash, K. L., Cvitanovic, C., Fulton, E. A., Halpern, B. S., Milner-Gulland, E. J., Watson, R. A., Blanchard, J., (2017). Planetary boundaries for a blue planet. Nat Ecol Evol 1, 1625–1634.
29. Ommer, R. E., (2018). Curiosity, interdisciplinarity, and giving back. ICES J Mar Sci 75, 1526–1535.
30. Pain, E., (2014). Better recognition for multidisciplinary research. Sci Mag. https://doi, teamwork and lessons for young scientists. ICES J Mar Sci 76, 28–40.
31. Pooley, S. P., Mendelsohn, J. A., Milner-Gulland, E. J., (2013). Hunting

down the chimera of multiple disciplinarity in conservation science. Conserv Biol 28, 22–32.

32. Rhoten, D., Parker, A., (2004). Risks and rewards of an interdisciplinary research path. Science 306, 2046.
33. Roy, E. D., Morzillo, A. T., Seijo, F., Reddy, S. M. W., Rhemtulla, J. M., Milder, J., Kuemmerle, C. T., Martin, S. L., (2013). The elusive pursuit of interdisciplinarity at the human-environment interface. Bioscience 63, 745–753.
34. Schwartz, M., Hiers, K., Davis, F. W., Garfin, G. M., Jackson, S. T., Teranso, A. J., Woodhouse, C. A., Morelli, T. L., Williamson, M. A., Brunson, M. W., (2017). Developing a translational ecology workforce. Front Ecol Environ 15, 587–596.
35. Stember, M., (1991). Advancing the social sciences through the interdisciplinary enterprise. Soc Sci J 28, 03623319.
36. Stephenson, R. L., Benson, A. J., Brooks, K., Degnbol, P., Dichmont, C. M., Kraan, M., Pascoe, S., Paul, S. D., Rindorf, A., Wiber, M., (2017). Practical steps toward integrating economic, social and institutional elements in fisheries policy and management. ICES J Mar Sci 74, 1981–1989.
37. Thiet, R., Dawson, A., Fagan, B., Hubbard, J., Lamonde, S., Meier, T., Settele, R., (2018). Advice to master's students for successfully navigating an interdisciplinary environmental studies program. Bull Ecol Soc Am 99, 314–326.
38. Thompson, S. A., Stephenson, R. L., Rose, G. A., Paul, S. D., (2019). Collaborative fisheries research: the Canadian Fisheries Research Network experience. Can J Fish Aquat Sci 76, 671–681.
39. Tress, B., Tres, G., Fry, G., (2006). From landscape research to landscape planning: aspects of integration, education and application. Springer, Berlin.
40. Turgeon, K., Hawkshaw, S. C. F., Dinning, K. M., Quinn, B. K., Edwards, D. N., Wor, C., Parlee, C. E., Debertin, A., Hawkshaw, M., Nelson, B. W., Zhang, F., Benestan, L., Angel, E., Morse BL, Mombourquette, D., (2017). Enhancing fisheries education and research through the Canadian Fisheries Research Network: a student perspective on interdisciplinarity, collaboration and inclusivity. FACETS 3, 1–17.
41. UN General Assembly (2015). Transforming our world: the 2030 agenda for sustainable development, 21 October 2015, A/RES/70/1. Available at: https://www.refworld, cross-disciplinarity, and sustainability science. Landsc Ecol 21, 1–4.

CHAPTER 8

Biogeographic Regions

CONTENTS

8.1. OVERVIEW

Biogeographic regions, also known as bioregions or ecoregions, are large geographic areas characterized by distinct ecological and biological features. These regions are defined by the distribution of species, ecosystems, and environmental conditions. Biogeographic regions play a crucial role in our understanding of biodiversity, conservation, and the distribution of life on Earth. In this overview of biogeographic regions, we will explore their significance, factors influencing their formation, and their role in ecological research and conservation.

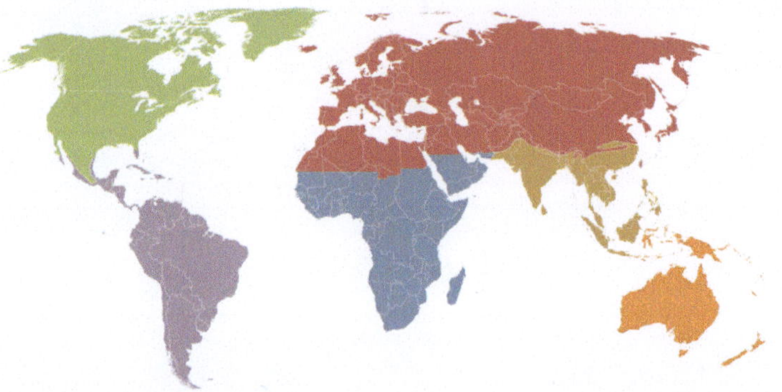

Figure 49. Biogeographic Regions

(Source: By carol – Ecozones and Image:BlankMap-World6, compact.svg by User:Lokal_Profil., CC BY-SA 3.0, https://commons.wikimedia.org/w/index.php?curid=3698005 , Creative Commons License)

Significance of Biogeographic Regions

Biodiversity Hotspots: Biogeographic regions are often associated with biodiversity hotspots—areas with exceptionally high levels of species richness and endemism. These regions are critical for conserving Earth's biological diversity.

- Conservation Prioritization: Understanding biogeographic regions helps conservationists identify priority areas for conservation efforts. Protecting these regions can have a significant impact on preserving global biodiversity.
- Research and Discovery: Biogeographic regions provide a framework for ecological research and the discovery of new species. Scientists study the unique assemblages of organisms

and ecosystems in these regions to better understand life's distribution patterns.

- Climate Change Impacts: Biogeographic regions can serve as indicators of climate change impacts. Shifts in the distribution of species and ecosystems within these regions can signal changes in temperature, precipitation, and habitat availability.

Factors Influencing Biogeographic Regions

- Geographic Barriers: Geographic features like mountains, oceans, and deserts can act as barriers, isolating populations and influencing the distribution of species. For example, the Andes Mountains in South America create distinct biogeographic regions on either side.
- Climate and Habitat: Climate and habitat conditions, such as temperature, rainfall patterns, and vegetation types, strongly influence the composition of biogeographic regions. These factors determine the suitability of an area for specific species.
- Historical Events: Historical events like glaciations, continental drift, and the formation of land bridges have shaped the distribution of species over geological time. For instance, the Bering Land Bridge connected Asia and North America during the last ice age, allowing species exchange.
- Evolutionary Processes: Evolutionary processes, including speciation and adaptive radiation, have contributed to the formation of unique biogeographic regions. Islands, for example, often have distinct biota due to isolated evolution.

Role in Ecological Research and Conservation

- Biogeographic Classification: Biogeographic regions are classified based on shared ecological characteristics, creating a framework for ecological research. This classification helps scientists study patterns of species distribution, evolution, and ecosystem dynamics.
- Conservation Planning: Conservation organizations use biogeographic regions to prioritize areas for protection and restoration efforts. By focusing on regions with high biodiversity and endemism, conservationists maximize the impact of their work.

Ecosystem Services: Biogeographic regions often provide essential ecosystem services, such as water purification, pollination, and climate regulation. Understanding these regions helps us appreciate the value of nature for human well-being.

S: As climate change alters the distribution of species and ecosystems, biogeographic regions can inform adaptive management strategies. Conservationists may need to consider assisted migration and habitat restoration to support species in changing conditions.

Biogeographic regions serve as critical tools for understanding the distribution of life on Earth and are central to biodiversity conservation and ecological research. They highlight the intricate relationship between geography, climate, evolution, and the diversity of species and ecosystems that make our planet unique.

8.2. BIOGEOGRAPHIC REALMS/ECOZONES

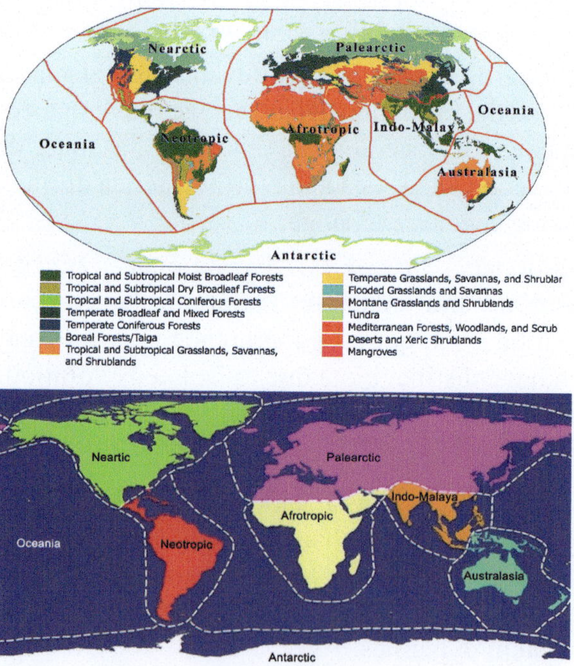

Figure 50. Ecozones

(Source: https://en.m.wikipedia.org/wiki/File:Ecozones_and_biomes_map.jpg, Creative Commons License)

Biogeographic realms, also known as ecozones, are large geographic regions characterized by distinct patterns of biodiversity, species distribution, and ecological communities. These realms are defined by the unique assemblages of organisms and environmental conditions that distinguish one region from another. In this overview of biogeographic realms, we will explore eight of the major realms on Earth: Nearctic, Palearctic, Afrotropical, Indomalaya, Australasian, Neotropical, Oceanian, and Antarctic.

Nearctic Realm

- Geographic Range: The Nearctic realm encompasses North America, including most of the United States and Canada.
- Key Features: This realm is characterized by diverse ecosystems, including temperate forests, grasslands, deserts, and tundra. Iconic species include American bison, grizzly bears, and bald eagles.

Palearctic Realm

- Geographic Range: The Palearctic realm extends across Europe, Asia, and Northern Africa.
- Key Features: This realm includes a wide range of habitats, from Siberian taiga to Mediterranean shrublands. Species like the Siberian tiger, Eurasian lynx, and European rabbit are found here.

Afrotropical Realm

- Geographic Range: The Afrotropical realm covers sub-Saharan Africa.
- Key Features: This realm is known for its incredible biodiversity, with savannas, rainforests, and deserts. Iconic species include African elephants, lions, and giraffes.

Indomalaya Realm

- Geographic Range: The Indomalaya realm includes South Asia and Southeast Asia.
- Key Features: This region is rich in rainforests, mangroves, and diverse ecosystems. Species like Bengal tigers, orangutans, and Asian elephants inhabit this realm.

Australasian Realm

- Geographic Range: The Australasian realm covers Australia, New Guinea, and nearby islands.
- Key Features: This is a region of unique and ancient biodiversity. Marsupials, such as kangaroos and koalas, are emblematic of this realm.

Neotropical Realm

- Geographic Range: The Neotropical realm includes Central and South America.
- Key Features: Rainforests, Andean mountains, and diverse ecosystems make this realm a biodiversity hotspot. Species like jaguars, macaws, and sloths are native to this region.

Oceanian Realm

- Geographic Range: The Oceanian realm encompasses the islands of the Pacific Ocean, including Hawaii and Polynesia.
- Key Features: This realm is characterized by volcanic islands and coral atolls. Unique species like the Hawaiian honeycreeper and marine life inhabit these islands.

Antarctic Realm

- Geographic Range: The Antarctic realm covers the continent of Antarctica and surrounding Southern Ocean.
- Key Features: This realm is defined by extreme cold, ice, and unique adaptations. Penguins, seals, and krill are iconic species found in this region.

Biogeographic realms play a crucial role in understanding the distribution of life on Earth and are fundamental to biodiversity conservation and ecological research. Each realm is home to its own set of species and ecosystems, shaped by its geographic history and environmental conditions. Studying these realms helps scientists unravel the complexity of life's diversity on our planet.

8.3. TERRESTRIAL REALMS

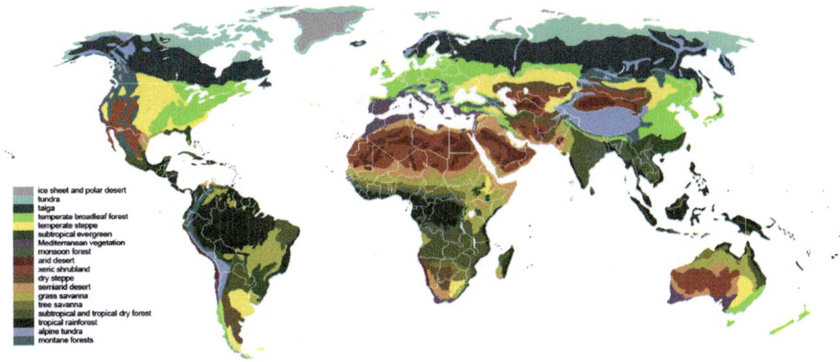

Figure 51. Terrestrial Realms

(Source: By Ville Koistinen (user Vzb83) – the blank world map in Commons and WSOY Iso karttakirja for the information, CC BY-SA 3.0, https://commons.wikimedia.org/w/index.php?curid=1700408, Creative Commons License)

Terrestrial realms are the major biogeographic regions on Earth's landmasses, each characterized by unique ecological communities, species distributions, and environmental conditions. These realms help us understand the patterns of biodiversity and the factors shaping life on land. Let's take a closer look at the terrestrial realms:

Nearctic Realm

- Geographic Range: North America, including most of the United States and Canada.
- Key Features: Diverse ecosystems such as temperate forests, grasslands, deserts, and tundra. Iconic species include American bison, grizzly bears, and bald eagles.

Palearctic Realm

- Geographic Range: Europe, Asia, and Northern Africa.
- Key Features: A wide range of habitats, from Siberian taiga to Mediterranean shrublands. Species like the Siberian tiger, Eurasian lynx, and European rabbit are found here.

Afrotropical Realm

- Geographic Range: Sub-Saharan Africa.
- Key Features: Incredible biodiversity with savannas, rainforests, and deserts. Iconic species include African elephants, lions, and giraffes.

Indomalaya Realm

- Geographic Range: South Asia and Southeast Asia.
- Key Features: Rich in rainforests, mangroves, and diverse ecosystems. Species like Bengal tigers, orangutans, and Asian elephants inhabit this realm.

Australasian Realm

- Geographic Range: Australia, New Guinea, and nearby islands.
- Key Features: A region of unique and ancient biodiversity. Marsupials, such as kangaroos and koalas, are emblematic of this realm.

Neotropical Realm

- Geographic Range: Central and South America.
- Key Features: Rainforests, Andean mountains, and diverse ecosystems make this realm a biodiversity hotspot. Species like jaguars, macaws, and sloths are native to this region.

These terrestrial realms are essential for understanding the distribution of life on land and play a significant role in biodiversity conservation and ecological research. Each realm has its own set of species and ecosystems, shaped by its geographic history and environmental conditions. Studying these realms helps scientists unravel the complexity of life's diversity on Earth's continents.

8.4. ECOREGIONS OF THE WORLD

Ecoregions are distinct geographic areas characterized by their unique ecosystems, species assemblages, and environmental conditions. They provide a finer-scale classification of terrestrial and aquatic environments, further refining our understanding of biodiversity and conservation priorities. Here, we explore some of the world's major ecoregions:

Tropical Rainforest Ecoregions

- Amazon Rainforest: Located in South America, the Amazon Rainforest is the world's largest tropical rainforest, known for its extraordinary biodiversity.
- Congo Basin: Situated in Central Africa, the Congo Basin houses extensive tropical rainforests and diverse wildlife, including gorillas and forest elephants.

Desert Ecoregions

- Sahara Desert: Spanning North Africa, the Sahara is the world's largest hot desert, characterized by its arid landscapes and extreme temperatures.
- Sonoran Desert: Found in North America, the Sonoran Desert is home to unique cacti species and adapted desert wildlife.

Mediterranean Ecoregions

- Mediterranean Basin: This ecoregion surrounds the Mediterranean Sea and features a Mediterranean climate, with distinctive flora and fauna.
- California Chaparral: Located in California, USA, this ecoregion comprises shrublands adapted to seasonal droughts and fires.

Grassland Ecoregions

- Pampas: Situated in South America, the Pampas is a fertile plain known for its grasslands and rich agricultural production.
- Savanna: Found in Africa, South America, and Australia, savannas are characterized by grasses and scattered trees, supporting diverse wildlife.

Arctic and Alpine Ecoregions

- Arctic Tundra: Extending across the Arctic regions of North America, Europe, and Asia, the tundra is a cold, treeless biome with unique adaptations.
- Alpine Tundra: High-elevation regions in mountain ranges worldwide, including the Alps and Rockies, feature alpine tundra ecosystems adapted to harsh conditions.

Coastal and Marine Ecoregions

- Great Barrier Reef: Off the coast of Australia, the Great Barrier Reef is the world's largest coral reef system, teeming with marine life.
- Galápagos Islands: This Pacific archipelago is renowned for its unique species and contributions to Charles Darwin's theory of evolution.

Boreal Forest Ecoregions

- Taiga: Extending across North America, Europe, and Asia, the taiga is characterized by coniferous forests and cold winters.
- Siberian Taiga: The vast Siberian taiga in Russia is one of the world's largest forested regions.

These ecoregions represent a fraction of Earth's ecological diversity. They provide essential frameworks for biodiversity conservation and ecosystem management, helping us prioritize and protect the planet's rich array of natural habitats and species. Understanding ecoregions enhances our ability to address global environmental challenges and promote sustainable stewardship of Earth's ecosystems.

8.5. REFERENCES

1. Aldecoa, R. Marín, I. (2013). Exploring the limits of community detection strategies in complex networks. Sci. Rep. *3,* 2216.
2. Drummond, A. J., Rambaut, A. (2007). Beast: Bayesian evolutionary analysis by sampling trees. BMC Evol. Biol. *7,* 214.
3. Edler, Rosvall, D. M., (2015). The infomap software package. Available from: URL http://www.mapequation.org.
4. Faurby, S. Svenning, J.-C. (2015). A species-level phylogeny of all extant and late quaternary extinct mammals using a novel heuristic-hierarchical bayesian approach. Mol. Phylogenet. Evol. *84,* 14–26.
5. Fitch, W. M. (1971). Toward defining the course of evolution: minimum change for a specific tree topology. Syst. Biol. *20,* 406–416.
6. Goldberg, E. E., Lancaster, L. T. Ree, R. H. (2011). Phylogenetic inference of reciprocal effects between geographic range evolution and diversification. Syst. Biol. *60,* 451–465.
7. Hargrove, W. W., Hoffman, F. M. Hessburg, P. F. (2006). Mapcurves: a quantitative method for comparing categorical maps. J. Geogr. Syst. *8,* 187–208.
8. Holt, B. G., Lessard, J.-P. Borregaard, M. K., Fritz, S. A. Araújo, M. B., Dimitrov, D. Fabre, P.-H. Graham, C. H., Graves, G. R. Jønsson, K. A., Nogués-Bravo, Wang, D., Whittaker, Z., R. J., Jeldså, Rahbek, J. C. (2013). An update of Wallace's zoogeographic regions of the world. *Science 339,* 74–78.
9. Kheirkhahzadeh, Lancichinetti, M., Rosvall, A. M. (2016). Efficient community detection of network flows for varying Markov times and bipartite networks. Phys. Rev. E *93,* 032309.
10. Kozak, K. H., & Wiens, J. (2006). Does niche conservatism promote speciation? A case study in North American Salamanders. Evolution *60,* 2604–2621.
11. Kreft, Jetz, H. W., (2010). A framework for delineating biogeographical regions based on species distributions. J. Biogeogr. *37,* 2029–2053.
12. Lancichinetti, Fortunato, A. S., (2009). Community detection algorithms: A comparative analysis. Phys. Rev. E *80,* 056117.
13. Landis, M. J., Matzke, N. J., Moore, B. R., & Huelsenbeck, J. P. (2013). Bayesian analysis of biogeography when the number of areas is large. Syst. Biol. *62,* 789–804.

14. Maldonado, Molina, C., C. I., Zizka, Persson, A., Taylor, C., C. M., Albán, Chilquillo, J., E. Rønsted, Antonelli, N. A., (2015). Estimating species diversity and distribution in the era of big data: to what extent can we trust public databases? Global Ecol. Biogeogr. *24,* 973–984.
15. Matzke, N. J. (2013). BioGeoBEARS: BioGeography with Bayesian (and Likelihood) Evolutionary Analysis in R Scripts. University of California, Berkeley, Berkeley, CA.
16. Matzke, N. J. (2014). Model selection in historical biogeography reveals that founder-event speciation is a crucial process in island clades. Syst. Biol. *63,* 951–970.
17. Meyer, Weigelt, C., Kreft, P. H., (2016). Multidimensional biases, gaps and uncertainties in global plant occurrence information. Ecol. Lett. *9,* 992–1006.
18. Morrone, J. J. (2006). Biogeographic areas and transition zones of Latin America and the Caribbean Islands based on panbiogeographic and cladistic analyses of the entomofauna. Annu. Rev. Entomol. *51,* 467–494.
19. Morrone, J. J. (2014). Biogeographical regionalisation of the neotropical region. Zootaxa 3*782,*1–110.
20. Oliveira, Brescovit, U., A. D., Santos, A. J. (2015). Delimiting areas of endemism through kernel interpolation. PLoS One *10,* e0116673.
21. Olson, D. M., Dinerstein, Wikramanayake, E., E. D., Burgess, N. D. Powell, G. V., Underwood, E. C. D'amico, J. A., Itoua, Strand, I., H. E., Morrison, J. C. Loucks, C. J., Allnutt, T. F. Ricketts, T. H., Kura, Lamoreux, Y., J. F., Wettengel, W. W. Hedao, Kassem, P., K. R. (2001). Terrestrial ecoregions of the world: a new map of life on earth a new global map of terrestrial ecoregions provides an innovative tool for conserving biodiversity. BioScience *51,* 933–938.
22. Ree, R. H. & Smith, S. A. (2008). Maximum likelihood inference of geographic range evolution by dispersal, local extinction, and cladogenesis. Syst. Biol. *57,* 4–14.
23. Rosvall, Bergstrom, M., C. T. (2008). Maps of random walks on complex networks reveal community structure. Proc. Natl. Acad. Sci. USA *105,* 1118–1123.
24. Töpel, Zizka, M., A. Calió, M. F., Scharn, Silvestro, R., Antonelli, D. A., (2016). Speciesgeocoder: fast categorization of species occurrences for analyses of biodiversity, biogeography, ecology, and evolution. Syst. Biol. http://sysbio.oxfordjournals, Creative Commons License).

CHAPTER 9

Interconnections in Nature

CONTENTS

9.1. ECOLOGY AS STUDY OF INTERCONNECTIONS IN NATURE

Ecology, often described as the science of interconnections in nature, is a multifaceted discipline that explores the relationships between living organisms and their environments. It is a branch of biology that delves deep into the intricate web of interactions that shape our planet's ecosystems. The essence of ecology lies in understanding not only the individual components of ecosystems but also the complex interplay between them. In this essay, we will explore the fundamental concepts of ecology, its significance in addressing contemporary environmental challenges, and its role in fostering a sustainable coexistence between humanity and the natural world.

At its core, ecology is concerned with the study of life and its environment. It seeks to answer fundamental questions about the natural world. How do species adapt to their surroundings? What are the mechanisms governing the abundance and distribution of organisms? How do ecosystems function and evolve over time? These questions underpin the foundation of ecology, guiding scientists in their quest to unravel the mysteries of the natural world.

Ecology operates at multiple levels of organization, ranging from individual organisms to entire ecosystems. Here are some key levels of ecological organization:

- *Individual Organisms:* At the lowest level, ecology explores the life history, behaviors, and adaptations of individual organisms. This is the realm of behavioral ecology, physiological ecology, and evolutionary ecology.
- *Populations:* Ecologists study groups of individuals of the same species, focusing on factors that influence population size, growth, and dynamics.
- *Communities:* Communities represent assemblages of multiple species living in the same area. Community ecology examines species interactions, including competition, predation, mutualism, and parasitism.
- *Ecosystems:* Ecosystem ecology delves into the flow of energy and the cycling of nutrients within ecological systems. It considers the abiotic (non-living) and biotic (living) components and their interactions.
- *Landscapes and Biomes:* On a larger scale, landscape ecology and biome ecology examine spatial patterns and processes across vast

regions, addressing issues of connectivity, habitat fragmentation, and biogeographic distributions.

Interactions in Ecology:

The heart of ecology lies in the study of interactions. These interactions can take myriad forms:

- *Predator-Prey Interactions:* The classic example of the predator-prey relationship, where predators consume prey, and regulates population dynamics in ecosystems.
- *Mutualistic Relationships:* Many species engage in mutually beneficial interactions, such as the pollination of flowers by insects or the nitrogen-fixing partnership between legumes and bacteria.
- *Competition:* Species often compete for limited resources, leading to resource partitioning and niche specialization to reduce competition.
- *Symbiosis:* Symbiotic relationships, including commensalism, mutualism, and parasitism, showcase the diversity of ecological interactions and their consequences for the organisms involved.

Ecology and Environmental Challenges:

In today's world, ecological knowledge is more critical than ever. Humanity faces pressing environmental challenges, including climate change, habitat loss, biodiversity loss, and pollution. Ecology provides essential tools for understanding and mitigating these challenges.

- *Climate Change:* Ecologists study the impacts of climate change on ecosystems and species distributions, helping us anticipate and adapt to shifting conditions.
- *Habitat Loss:* Understanding the consequences of habitat destruction informs conservation efforts and strategies for habitat restoration.
- *Biodiversity Conservation:* Ecology is fundamental to the preservation of biodiversity, as it uncovers the intricacies of species interactions and the importance of preserving diverse ecosystems.
- *Pollution and Toxins:* Ecotoxicology examines the effects of

pollutants on organisms and ecosystems, guiding regulations and pollution management.

Fostering Sustainable Coexistence:

Ecology serves as a guiding light for achieving a sustainable balance between human activities and the natural world. It emphasizes the importance of maintaining healthy ecosystems to ensure our own well-being. By recognizing that all living organisms are interconnected, ecology encourages a holistic perspective on environmental management.

- *Ecosystem-Based Management:* Ecological principles guide ecosystem-based management approaches, ensuring that human activities consider the health and resilience of ecosystems.
- *Conservation Strategies:* Conservation efforts leverage ecological insights to prioritize protected areas, restore habitats, and save endangered species.
- *Sustainable Agriculture:* Ecological farming practices promote sustainability by working with natural processes rather than against them.
- *Climate Mitigation:* Ecological restoration and reforestation efforts play a role in mitigating climate change by sequestering carbon and preserving natural carbon sinks.

In summary, ecology, as the science of interconnections in nature, embodies the profound idea that every living organism is part of a larger ecological tapestry. It demonstrates that our actions have consequences beyond what meets the eye, affecting the intricate balance of ecosystems. As we navigate an era of unprecedented environmental challenges, ecology's role in understanding and addressing these issues is paramount. It empowers us to make informed decisions, implement sustainable practices, and strive for harmony between human activities and the natural world. In doing so, it offers hope for a future where both humanity and the planet can thrive together.

9.2. REFERENCES

1. v Alcamo, J., (ed.) (1990). The RAINS Model of Acidification: Science and Strategies in Europe. Kluwer Academic, Dordrecht.
2. Alcamo, J., Kreileman, G. J. J., Krol, M. S., & Zuidema, G. (1994). Modeling the global society-biosphere-climate system. Part 1, Model description and testing. Water, Air and Soil Pollution 76(March):1–35.
3. Alcamo, J., Krol, M., & Leemans, R. (1995). Stabilizing Greenhouse Gases: Global and Regional Consequences. National Institute of Public Health and the Environment (RIVM), Bilthoven, The Netherlands.
4. Allen, J. C., & Barnes, D. F. (1985). The causes of deforestation in developing countries. Annals of the Association of American Geographers *75*(2), 163–184.
5. Appendini,, K., & Liverman, D. M. (1994). Agricultural policy and climate change in Mexico. Food Policy *19*(2), 149–164.
6. Arrow, K., Bolin, B., Costanza, R., Dasgupta, P., Folke, C., C. S. Holling, Jansson, B., S. Levin, Maler, K., C. Perrings, & Pimentel, D. (1995). Economic growth, carrying capacity and the environment. *Science* 268(5210):520–521.
7. Arthur, W. B. (1994). Increasing Returns and Path Dependence in the Economy. University of Michigan Press, Ann Arbor.
8. Axelrod, R. M. (1984). The Evolution of Cooperation. Basic Books, New York.
9. Ayres, R. U., & Ayres, L. (1996). Industrial Ecology: Towards Closing the Materials Cycle. E. Elgar, Brookfield, Vt.
10. Bailey, C., Jentoft, S., & Sinclair, P., (eds.) (1996). Aquacultural Development: Social Dimensions of an Emerging Industry. Westview Press, Boulder, Colo.
11. Balee, W. L. (1994). Footprints of The Forest: Ka'apor Ethnobotany—The Historical Ecology of Plant Utilization by an Amazonian People. Columbia University Press , New York.
12. Baland, J-M., & Platteau, J-P. (1996). Halting Degradation of Natural Resources: Is There a Role for Rural Communities? Clarendon Press, Oxford.
13. Barke, R. P., Jenkins-Smith, H., & Slovic, P. (1997). Risk perceptions of men and women scientists. Social Science Quarterly *78*(1), 167–176.
14. Barrett, S. (1996). Building property rights for transboundary

resources. pp. 265–284 in Rights to Nature, S. Hanna, ed. Island Press, Washington, D.C.

15. Baumol, W. J., & Oates, W. E. (1988). The Theory of Environmental Policy, 2d ed. Cambridge University Press, Cambridge, U.K.
16. Beamish, R. J., ed. (1995). Climate change and northern fish populations. Special Publication 121. Canadian Journal of Fisheries and Aquatic Sciences.
17. Bentham, G. (1993). Depletion of the ozone layer: Consequences for non-infectious human diseases. Parasitology *106,* S39–S46.
18. Berkes, F., Feeny, D., McCay, B. J., & Acheson, J. M. (1989). The benefit of the commons. Nature 340(July 13):91–93.
19. Bijlsma, L. (1996). Coastal zones and small islands. pp. 289–234 in Climate Change 1*995,* Impacts, Adaptation and Mitigation of Climate Change: Scientific and Technical Analyses, R. T. Watson et al., (eds.) Cambridge University Press, Cambridge, U.K.
20. Black, R. (1994). Forced migration and environmental change: The impact of refugees on host environments. Journal of Environmental Management *42*(3)*,* 261–278.
21. Blaikie, P. M. (1994). At Risk: Natural Hazards, People's Vulnerability, & Disasters. Routledge, London, New York.
22. Blaikie, P., & Brokfield, H. C. (1987). Land Degradation and Society. Methuen, London.
23. Blinder, A. (1987). Hard Heads, Soft Hearts: Tough-Minded Economics for a Just Society. Addison-Wesley, Reading, Mass.
24. Bohi, D. (1981). Analyzing Demand Behavior: A Study of Energy Elasticities. Johns Hopkins University Press, Baltimore, Md.
25. Bohle, H. G., Downing, T. E., & Watts, M. J. (1994). Climate change and social vulnerability: Toward a sociology and geography of food insecurity. Global Environmental Change *4*(1)*,*37–48.
26. Bongaarts, J. (1994). The impact of population policies: Comment. Population and Development Review *20*(3)*,* 616–620.
27. Boserup, E. (1965). The Conditions of Agricultural Growth. Aldine, Chicago.
28. Boserup, E. (1981). Population and Technological Change: A Study of

Long-Term Trends. University of Chicago Press, Chicago.

29. Brander, K. (1996). Effects of climate change on cod (Gadus murhua) stocks. pp. 255–278 in Global Warming: Implications for Freshwater and Marine Fish, C. M. Wood, & G, D. McDonald, (eds.) Society for Experimental Biology Seminar Series 61. Cambridge University Press, Cambridge, U.K.
30. Brewer, G. D. (1986). Methods for synthesis: Policy exercises. pp. 455–473 in Sustainable Development of the Biosphere, W. C. Clark, & E, R. Munn, (eds.) Cambridge University Press, Cambridge, U.K.
31. Brody, J. G., & Rosen, R. A. (1994). Apples and oranges: Using multi-attribute utility analysis in a collaborative process to address value conflicts in electric facility siting. NRRI Quarterly Bulletin (National Regulatory Research Institute) *15*(4), 629–643.
32. Bromley, D., ed. (1992). Making the Commons Work: Theory, Practice, & Policy. ICS Press, San Francisco.
33. Brondizio, E., Moran, E. F., Mausel, P., & Wu, R. (1996). Land cover in the Amazon estuary: Linking of TM with botanical and historical data. Photogrammatic Engineering and Remote Sensing *62,* 921–930.
34. Brooks, H. (1986). A typology of surprises in technology, institutions and development. In Sustainable Development of the Biosphere, W. C. Clark, & E, R. Munn, (eds.) Cambridge University Press, New York.
35. Brulle, R. J. (1995). Environmental discourse and social movement organizations: A historical and rhetorical perspective on the development of, U.S. environmental organizations. Sociological Inquiry *66,* 58–83.
36. Burton, I., Kates, R. W., & White, G. F. (1993). The Environment as Hazard, 2d ed. Oxford University Press, New York.
37. Cane, M. A., Eshel, G., & Buckland, R. W. (1994). Forecasting Zimbabwean maize yields using eastern equatorial Pacific sea surface temperatures. Nature *370,* 204–205.
38. Caulfield, C. (1985). The rain forests. The New Yorker 60(Jan. 14):41–83.
39. Chayes, A., & Chayes, A. H. (1995). The New Sovereignty: Compliance with International Regulatory Agreements. Harvard University Press, Cambridge, Mass.
40. Cohen, S. (1996). Mackenzie Basin Impact Study Final Report. Environment Canada, Ottawa, Canada.

41. Colwell, R. R. (1996). Global climate and infectious disease: The cholera paradigm. *Science* 274(5295):2025–2031.
42. Cordell, J. ed. (1989). A Sea of Small Boats. Cultural Survival Report 26. Cultural Survival, Inc., Cambridge, Mass.
43. Cowen, D. J., & Jensen, J. R. (1998). Urban change and development in southeastern United States. In People and Pixels: Linking Remote Sensing and Social Science. D. M. Liverman et al., (eds.) National Academy Press, Washington, D.C.
44. Crocker, D. A., & Linden, T., (eds.) (1998). Ethics of Consumption: The Good Life, Justice, & Global Stewardship. Rowman and Littlefield, Lanham, Md.
45. Cropper, M. L., & Oates, W. E. (1992). Environmental economics: A survey. Journal of Economic Literature *30,* 675–704.
46. Cropper, M. L., Aydede, S. K., & Portney, P. R. (1994). Preferences for life saving programs: How the public discounts time and age. Journal of Risk and Uncertainty *8,* 243–246.
47. Crosby, A. W. (1972). The Columbian Exchange: Biological and Cultural Consequences of (1492). Greenwood Publishing Co., Westport, Conn.
48. Daedalus. (1996). The liberation of the environment. 125(3), 1–253.
49. Dale, V. H. (1994). Effects of Land-Use Change on Atmospheric CO2 Concentrations: South and Southeast Asia as a Case Study. Springer-Verlag, New York.
50. De Bardeleben, J. (1985). The Environment and Marxism-Leninism: The Soviet and Eastern German Experience. Westview Press, Boulder, Colo.
51. Deudney, D. (1990). The case against linking environmental degradation and national security. Journal of International Studies *19*(3), 461–476.
52. Diamond, P. A., & Hausman, J. A. (1994). Contingent valuation: Is some number better than no number? Journal of Economic Perspectives *8,* 45–64.
53. Dicken, P. (1992). Europe 1992 and strategic change in the international automobile industry. Special issue: Industrial Restructuring and Continental Trade Blocs. Environment & Planning *24*(1), 11–31.
54. Dietz, T., & Rosa, E. A. (1997). Effects of population and affluence on CO2 emissions. *Proceedings of the National Academy of Sciences 94*(1), 175–179.

55. Dietz, T., & Stern, P. C. (1998). Science, values, and biodiversity. *Bioscience* 48(June):441–444.
56. Dietz, T., Stern, P. C., & Rycroft, R. W. (1989). Definitions of conflict and the legitimation of resources: The case of environmental risk. Sociological Forum *4,* 47–70.
57. Dillman, D., Rosa, E. A., & Dillman, J. J. (1983). Lifestyle and home energy conservation in the, U.S. Journal of Economic Psychology *3,* 299–315.
58. Downing, T. E., ed. (1995). Climate Change and World Food Security. Springer-Verlag, Heidelberg.
59. Drabek, T. E. (1986). Human Systems' Responses to Disaster. Springer-Verlag, New York.
60. Dunlap, R. E. (1992). Trends in public opinion toward environmental issues: 1965–1990. pp. 89–116 in American Environmentalism: The, U.S. Environmental Movement, 1970–1990, R. E. Dunlap, & G, A. Mertig, (eds.) Taylor and Francis, Washington, D.C.
61. Dunlap, R. E., & G, A. Mertig, (eds.) (1992). American Environmentalism: The, U.S. Environmental Movement, 1970–1990. Taylor and Francis, Washington, D.C.
62. Dunlap, R. E., Gallup, G. H., & Gallup, A. M. (1993). Of global concern: Results of the health of the planet survey. Environment *35*(9), 6–22.
63. Easterling, W. E., III, P. R. Crosson, N. J. Rosenberg, M. S. McKenney, L. A. Katz, & Lemon, K. M. (1993). Agricultural impacts of and responses to climate change in the Missouri-Iowa-Nebraska-Kansas (MINK) region. Climatic Change 24(1–2):23–39.
64. Edmonds, J. A., Wise, M. A., & MacCracken, C. (1994). Advanced Energy Technologies and Climate Change: An Analysis Using the Global Change Assessment Model (GCAM). PNL-9798, UC-402. Pacific Northwest Laboratory, Richland, Wash.
65. Edwards, W., & Newman, J. R. (1982). Multi-attribute evaluation. Quantitative Applications in the Social Sciences Series, Paper No. 26. Sage, New York.
66. El-Hinnawi, E. (1985). Environmental Refugees. United Nations Development Program, New York.
67. Ellis, R. D. (1993). Quantifying distributive justice: An approach to

environmental and risk-related public policy. Policy Sciences *26,* 99–103.

68. Entwistle, B., Rindfuss, R. R., & Walsh, S. J. (1998). Land use/land cover and population dynamics, Nang Rong, Thailand. In People and Pixels: Linking Remote Sensing and Social Science. D. M. Liverman et al., (eds.) National Academy Press, Washington, D.C.
69. European Commission. (1995). Global Environmental Change and Sustainable Development in Europe, Jager, J., A. Liberatore, & Grundlach, K., (eds.) DG X11. European Commission, Luxembourg.
70. Everett, J. T. (1996). Fisheries. pp. 511–537 in Climate Change 1*995,* Impacts, Adaptation and Mitigation of Climate Change: Scientific and Technical Analyses, R. T. Watson et al., (eds.) Cambridge University Press, Cambridge, U.K.
71. Food and Agriculture Organization. (1997). Online database: Hyperlink http://apps.fao, M. (1990). Perspectives on consumer culture: The sociology of consumption. Sociology *24*(1)*,* 5–22.
72. Feeny, D., Berkes, F., McCay, B., & Acheson, J. (1990). The tragedy of the commons: Twenty-two years later. Human Ecology *18,* 1–19.
73. Finlayson, A. C., & McCay, B. J. (1998). Crossing the threshold of ecosystem resilience: The commercial extinction of northern cod. pp. 311–338 in Linking Social and Ecological Systems: Institutional Learning for Resilience, C. Folke, & Berkes, F., (eds.) Cambridge University Press, Cambridge, U.K.
74. Fischhoff, B., Lichtenstein, S., Slovic, P., Derby, S., & Keeney, R. (1981). Acceptable Risk. Cambridge University Press, New York.
75. Fischhoff, B., Watson, S. R., & Hope, C. (1984). Defining risk. Policy Sciences *17,* 123–139.
76. Flynn, J., Slovic, P., & Mertz, C. K. (1994). Gender, race, and perception of environmental health risks. Risk Analysis *14,* 1101–1108.
77. Folke, C., & Kautsky, N. (1996). The ecological footprint concept for sustainable seafood production. Unpublished paper presented at the Conference on Ecosystem Management for Sustainable Marine Fisheries, Monterey, Calif, Feb. 19–23.
78. Foresta, R. A. (1992). Amazonia and the politics of geopolitics. The Geographical Review *82*(2)*,* 128–142.
79. Gamson, W. E., & Modigliani, A. (1989). Media discourse and public opinion on nuclear power: A constructionist approach. The American

Journal of Sociology *95*(1), 1–37.

80. Gardner, G. T., & Stern, P. C. (1996). Environmental Problems and Human Behavior. Allyn and Bacon, Needham Heights, Mass.
81. Gillis, M., & Repetto, R. C. (1988). Public Policies and the Misuse of Forest Resources. Cambridge University Press, New York.
82. Glacken, C. J. (1967). Traces on the Rhodian Shore. University of California Press, Berkeley.
83. Gleick, P. (1989). Climate change and international politics. Ambio *18*(6), 333–339.
84. Goodwin, N. R., Ackerman, F., & Kiron, D., (eds.) (1997). The Consumer Society. Island Press, Washington, D.C.
85. Greenpeace International. (1994). The climate time bomb. http://www.greenpeace, R., Lichtenstein, S., & Slovic, P. (1993). Valuing environmental resources: A constructive approach. Journal of Risk and Uncertainty *7,* 177–197.
86. Griliches, Z. (1957). Hybrid corn: An exploration in the economics of technological change. Econometrica *25,* 501–522.
87. Grobecker, A. J., Coroniti, S. C., & H, R. Cannon, Jr. (1974). The Report of Findings: The Effects of Stratospheric Pollution by Aircraft. DOT-TST-75–50. U.S. Department of Transportation, Climatic Impact Assessment Program. National Technical Information Service, Springfield, Va.
88. Grossman, G., & Krueger, A. (1995). Economic growth and the environment. Quarterly Journal of Economics 11*0*(2),353–377.
89. Groves, T., Radner, R., & Reiter, S. (1987). Information Incentives, & Economic Mechanisms: Essays in Honor of Leonid Hurwicz. University of Minnesota Press, Minneapolis.
90. Grubb, M., Chapuis, T., & Duong, M. H. (1995). The economics of changing course: Implications of adaptability and inertia for optimal climate policy. Energy Policy 23(April–May):417–432.
91. Grubler, A. (1996). Time for a change: On the patterns of diffusion of innovation. Daedalus 12*5*(3),19–42.
92. Guyer, J. I., & Lambin, E. F. (1993). Land use in an urban hinterland: Ethnography and remote sensing in the study of African intensification. American Anthropologist *95*(4), 839.
93. Haas, P. M. (1990). Saving the Mediterranean: The Politics of

International Environmental Cooperation. Columbia University Press, New York.

94. Haas, P. M. (1993). Protecting the Baltic and North seas. In Institutions for the Earth: Sources of Effective International Environmental Protection, Haas, P., R. Keohane, & A, M. Levy, (eds.) MIT Press, Cambridge, Mass.
95. Haas, P. M., Keohane, R. O., & A, M. Levy, (eds.) (1993). Institutions for the Earth: Sources of Effective International Environmental Protection. MIT Press, Cambridge, Mass.
96. Hammitt, J. K., Lempert, R. J., & Schlesinger, M. E. (1992). A sequential-decision strategy for abating climate change. Nature 357(6376):315–18.
97. Hanemann, W. M. (1994). Valuing the environment through contingent valuation. Journal of Economic Perspectives *8,* 19–43.
98. Hanna, S., ed.. (1996). Rights to Nature. Island Press, Washington, D.C.
99. Hanna, S., & Munasinghe, M., (eds.) (1995a). Property Rights and the Environment: Social and Ecological Issues. The World Bank and the Beijer Institute for Ecological Economics, Washington, D.C.
100. Hanna, S., & Munasinghe, M., (eds.) (1995b). Property Rights in a Social and Ecological Context: Case Studies and Design Applications. The World Bank and the Beijer Institute for Ecological Economics, Washington, D.C.
101. Hardin, G. (1968). The tragedy of the commons. *Science 162,* 1243–1248.
102. Hayami, Y., & Ruttan, V. (1985). Agricultural Development: An International Perspective. Johns Hopkins University Press, Baltimore.
103. Hayflick, L. (1994). How and Why We Age. Ballantine Books, New York.
104. Hecht, S. B., & Cockburn, A. (1989). Fate of the Forest: Developers, Destroyers and Defenders of the Amazon. Verso, New York.
105. Hempel, L. C. (1996). Environmental Governance: The Global Challenge. Island Press, Washington, D.C.
106. Herman, R., Ardekani, S. A., & Ausubel, J. H. (1989). Dematerialization. pp. 50–69 in Technology and Environment. National Academy Press, Washington, D.C.

107. Holtz-Eakin, D., & Selden, T. M. (1995). Stoking the fires? CO2 emissions and economic growth. Journal of Public Economics *57,* 85–101.
108. Homer-Dixon, T. F. (1991). On the threshold: Environmental changes as causes of acute conflict. International Security *16*(2), 76.
109. Homer-Dixon, T. F., & Levy, M. A. (1995). Environment and security. International Security *20*(3), 189.
110. Homer-Dixon, T. F., J.H Boutwell, & Rathjens, G. W. (1993). Environmental change and violent conflict. Scientific American 26*8*(2),38–45.
111. Hope, C., Anderson, J., & Wenman, P. (1993). Policy analysis of the greenhouse effect: An application of the PAGE model. Energy Policy 21(March):327–338.
112. Hourcade, J., & Chapuis, T. (1995). No-regret potentials and technical innovation: A viability approach to integrated assessment of climate change. Energy Policy 23(April/May):433–446.
113. Hugo, G. (1996). Environmental concerns and international migration. International Migration Review *30*(1), 105–131.
114. Hutchings, J. A. (1996). Spatial and temporal variation in the density of Northern Cod and a review of hypotheses for the stock's collapse. Canadian Journal of Fisheries and Aquatic Sciences *53,* 943–962.
115. Inglehart, R. (1990). Culture Shift in Advanced Industrial Society. Princeton University Press, Princeton, N. J.
116. Inglehart, R. (1995). Public support for environmental protection: Objective problems and subjective values in 43 societies. Political Science and Politics *15,* 57–71.
117. Inglehart, R. (1997). Modernization and Postmodernization: Cultural, Economic and Political Change in 43 Societies. Princeton University Press, Princeton, N. J.
118. Intergovernmental Panel on Climate Change. (1996a). Climate Change 1*995,* Impacts, Adaptation and Mitigation of Climate Change: Scientific and Technical Analyses , R. T. Watson et al., (eds.) Cambridge University Press, Cambridge, U.K.
119. Intergovernmental Panel on Climate Change. (1996b). Economic and Social Dimensions of Climate Change 1995, J. Bruce et al., (eds.) Cambridge University Press, Cambridge, U.K.
120. International Energy Agency. (1997). Indicators of Energy Use and

Efficiency: Understanding the Link Between Energy and Human Activity. International Energy Agency, Paris.

121. Jacobson, H. K., & Weiss, E. B. (1990). Implementing and complying with international environmental accords: A framework for research. Unpublished paper presented at the annual meeting of the American Political Science Association, San Francisco.
122. Jasanoff, S. (1986). Risk Management and Political Culture. Russell Sage Foundation, New York.
123. Johnson, P., & Beaulieu, A., (eds.) (1996). The Environment and NAFTA: Understanding and Implementing the New Continental Law. Island Press, Washington, D.C.
124. Johnston, R. J., Taylor, P., & Watts, M. (1995). Geographies of Global Change: Remapping the World in the Late Twentieth Century. Blackwell, Oxford.
125. Jones, R. E., & Dunlap, R. E. (1992). The social bases of environmental concern: Have they changed over time? Rural Sociology *57,* 28–47.
126. Kahneman, D., Ritov, I., Jacowitz, K. E., & Grant, P. (1993). Stated willingness to pay for public goods: A psychological perspective. Psychological Science *4,* 310–315.
127. Kalkstein, L. S. (1993). Health and climate change—direct impacts in cities. The Lancet *342,* 1397–1399.
128. Kalkstein, L. S. (1995). Cities could face killer heat waves. *Science 267,* 958.
129. Kalkstein, L. S., & Tan, G. (1995). Climate change and human health: International implications. pp. 124–145 in As Climate Changes: International Impacts and Implications, K. M. Strzepek, & B, J. Smith, (eds.) Cambridge University Press, New York.
130. Kaplan, R. D. (1994). The coming anarchy: How scarcity, crime, overpopulation, tribalism, and disease are rapidly destroying the social fabric of our planet. The Atlantic Monthly 27*3*(2)*,*44–64.
131. Kates, R. W., Ausubel, J. H., & Berberian, M., (eds.) (1985). Climate Impact Assessment: Studies of the Interaction of Climate and Society. Published on behalf of the Scientific Committee on Problems of the Environment of the International Council for Scientific Unions. John Wiley & Sons, New York.
132. Keeney, R. L., & Raiffa, H. (1976). Decisions with Multiple Objectives: Preferences and Value Tradeoffs. John Wiley & Sons, New York.

133. Kempton, W. (1991). Lay perspectives on global climate change. Global Environmental Change: Human and Policy Dimensions *1,* 183–208.
134. Kempton, W. (1993). Will public environmental concern lead to action on global warming? Annual Review of Energy and Environment *18,* 217–245.
135. Kempton, W., Boster, J. S., & Hartley, J. A. (1995). Environmental Values in American Culture. MIT Press, Cambridge, Mass.
136. Keohane, R. O. With the collaboration of, E. Ostrom, & McGinnis, M. (1993). Linking local and global commons: Monitoring, sanctioning, and theories of self-organization in common pool resources and international regimes. pp. 1–15 in Proceedings of a Conference on Linking Local and Global Commons, R. O. Keohane, M. McGinnis, & Ostrom, E., (eds.) Center for International Affairs, Harvard University, Cambridge, Mass.
137. Keohane, R. O., & A, M. Levy, (eds.) (1996). Institutions for Environmental Aid. MIT Press, Cambridge, Mass.
138. Keohane, R. O., & Ostrom, E., (eds.) (1995). Local Commons and Global Interdependence: Heterogeneity and Cooperation in Two Domains. Sage, London.
139. Kinzig, A. P., & Socolow, R. H. (1994). Human impacts on the nitrogen cycle. Physics Today 47(11):24–31.
140. Kolstad, C. D. (1996). Learning and stock effects in environmental regulation: The case of greenhouse gas emissions. Journal of Environmental Economics and Management *31,* 1–18.
141. Krasner, S., ed.. (1983). International Regimes. Cornell University Press, Ithaca.
142. Krimsky, S., & Golding, D. (1992). Social Theories of Risk. Praeger, Westport, Conn.
143. Lamb, H. H. (1995). Climate, History and the Modern World. Routledge, New York.
144. Lave, L. B., & Dowlatabadi, H. (1993). Climate change: The effect of personal beliefs and scientific uncertainty. Environmental Science and Technology 27(10):1962–1972.
145. Leigh, J. P. (1989). Compensating wages for job-related death: The opposing arguments. Journal of Economic Issues *23,* 823–839.
146. Levine, B. L. (1986). The tragedy of the commons and the comedy

of community: The commons in history. Journal of Community Psychology *14,* 81–99.

147. Levy, M. A., Young, O. R., & Zuern, M. (1995). The study of international regimes. European Journal of International Relations *1,* 267–330.
148. Litfin, K. T. (1994). Ozone Discourses: Science and Politics in Global Environmental Cooperation. Columbia University Press, New York.
149. Liverman, D. M. (1992). The regional impacts of global warming in Mexico: Uncertainty, vulnerability and response. pp. 44–68 in The Regions and Global Warming, J. Schmandt, & Clarkson, J., (eds.) Oxford University Press, Oxford, England.
150. Liverman, D. M. (1994a). Environment and security in Mexico. In Mexico: In Search of Security, B. M. Bagley, & S. Aguayo Quezada, (eds.) Transaction Publishers, New Brunswick, N. J.
151. Liverman, D. M. (1994b). Vulnerability to global environmental change. pp. 326–342 in Environmental Risks and Hazards, S. Cutter, ed. Prentice-Hall, Englewood Cliffs, N. J.
152. Löfstedt, R. E. (1992). Swedish lay perspectives on global climate change. Energy and Environment *3,* 161–175.
153. Löfstedt, R. E. (1995). Lay perspectives concerning global climate change in Vienna, Austria. Energy and Environment *4,* 140–154.
154. Lonergan, S. C., & Kavanagh, B. (1991). Climate change, water resources and security in the Middle East. Global Environmental Change (Sept.):272–290.
155. Ludwig, D., Hilborn, R., & Walters, C. (1993). Uncertainty, resource exploitation, and conservation: Lessons from history. *Science* 260(April 2):17.
156. Lutzenhiser, L. (1993). Social and behavioral aspects of energy use. Annual Review of Energy and the Environment *18,* 247–289.
157. Lutzenhiser, L. (1997). Social structure, culture, and technology: Modeling the driving forces of household energy consumption. pp. 77–91 in Environmentally Significant Consumption: Research Directions, P. C. Stern et al., (eds.) National Academy Press, Washington, D.C.
158. Manne, A. S., Mendelsohn, R., & Richels, R. G. (1993). MERGE: A model for evaluating regional and global effects of GHG reduction policies. Energy Policy *23*(1), 17–34.
159. March, J. G., & Olsen, J. P. (1989). Rediscovering Institutions: The

Organizational Basis of Politics. The Free Press, New York.

160. Mausel, P., Y. Wu., Li, Y., Moran, E., & Brondizio, E. (1993). Spectral identification of successional stages following deforestation in the Amazon. Geocarto International *8,* 11.

161. Mazur, A., & Lee, J. (1993). Sounding the global alarm: Environmental issues in the, U.S. national news. Social Studies of Science *23,* 681–720.

162. McCay, B. J. (1995). Social and ecological implications of ITQs: An overview. Ocean and Coastal Management 28(1–3):3–22.

163. McCay, B. J., & Acheson, J. M. (1987). The Question of the Commons: The Culture and Ecology of Communal Resources. University of Arizona Press, Tucson.

164. McCay, B. J., & Jentoft, S. (1996). Unvertrautes Gelande: Gemeineigentum Unter der Sozialwissenschaftlichen Lupe (Uncommon Ground: Perspectives on Common Property). Kölner Zeitschrift für Soziologie und Sozialpsychologie (Cologne Journal of Sociology and Social Psychology) 36(Fall):272–291.

165. McGoodwin, J. R. (1990). Crisis in the World's Fisheries: People, Problems and Politics. Stanford University Press, Stanford, Calif.

166. McKean, M. A. (1992). Success on the commons: A comparative examination of institutions for common property resource management. Journal of Theoretical Politics *4*(3),247–281.

167. McMichael, A. J. (1996). Human population health. pp. 561–584 in Climate Change 1*995,* Impacts, Adaptation and Mitigation of Climate Change: Scientific and Technical Analyses, R. T. Watson et al., (eds.) Cambridge University Press, Cambridge, U.K.

168. McNeill, J. R. (1992). The Mountains of the Mediterranean World: An Environmental History. Cambridge University Press, New York.

169. Meltzoff, S. K., & LiPuma, E. (1986). The social and political economy of coastal zone management: Shrimp mariculture in Ecuador. Coastal Zone Management Journal *14*(4), 349–380.

170. Mendelsohn, R. O., & Neumann, J. E. (1998). The Impact of Climate Change on the United States Economy. Cambridge University Press, New York.

171. Mileti, D. S., Drabek, T. E., & Haas, J. E. (1975). Human Systems in Extreme Environments: A Sociological Perspective. University of Colorado Institute of Behavioral Science, Boulder.

172. Mileti, D. S., Darlington, J. D., Passerini, E., Forrest, B. C., & Myers, M. F. (1995). Toward an integration of natural hazards and sustainability. The Environmental Professional *17*(2), 117–126.
173. Mintzer, I. S. (1987). A Matter of Degrees: The Potential for Controlling the Greenhouse Effect. World Resources Institute, Washington, D.C.
174. Mitchell, R. B. (1994). Intentional Oil Pollution at Sea: Environmental Policy and Treaty Compliance. MIT Press, Cambridge, Mass.
175. Mitchell, R. C., & Carson, R. T. (1989). Using Surveys to Value Public Goods: The Contingent Valuation Method. Resources for the Future, Washington, D.C.
176. Moran, E. F. (1981). Developing the Amazon. Indiana University Press, Bloomington.
177. Moran, E. F. (1995). The Comparative Analysis of Human Societies: Toward Common Standards for Data Collection and Reporting. L. Rienner Publishers, Boulder, Colo.
178. Moran, E. F., & Brondizio, E. (1998). Land-use change after deforestation in Amazonia. In People and Pixels: Linking Remote Sensing and Social Science, D. M. Liverman et al., (eds.) National Academy Press, Washington, D.C.
179. Moran, E. F., Brondizio, E., Mausel, P., & Wu, Y. (1994). Integrating Amazonian vegetation, land use and satellite data. BioScience *44*(5), 329–338.
180. Moran, E. F., Packer, A., Brondizio, E., & Tucker, J. (1996). Restoration of vegetation cover in the eastern Amazon. Ecological Economics *18,* 41–54.
181. Morgan, M. G., & Dowlatabadi, H. (1996). Learning from integrated assessment. Climatic Change 34(Dec.):337–368.
182. Myers, N. (1993). Ultimate Security: The Environmental Basis of Political Stability. W. W. Norton, New York.
183. Nakicenovic, N. (1996). Freeing energy from carbon. Daedalus *125*(3),95–112.
184. National Academy of Engineering. (1994). The Greening of Industrial Ecosystems, B. Allenby, & J, D. Richards, (eds.) National Academy Press, Washington, D.C.
185. National Academy of Engineering. (1989). Technology and Environment. J. Ausubel, & Sladovich, H., (eds.) National Academy Press, Washington, D.C.

186. National Acid Precipitation Assessment Program. (1991). The Oversight Review Board of the National Acid Precipitation Assessment Program. The Experience and Legacy of NAPAP. NAPAP, Oversight Review Board, Washington, D.C.
187. National Research Council. (1984a). Energy Use: The Human Dimension, P. C. Stern, & Aronson, E., (eds.) Freeman: New York.
188. National Research Council. (1984b). Improving Energy Demand Analysis, P. C. Stern, ed. National Academy Press, Washington, D.C.
189. National Research Council. (1988). The human dimensions of global environmental change. pp. 134–200 in Toward an Understanding of Global Change: Initial Priorities for, U.S. Contribution to the International Geosphere-Biosphere Program. National Academy Press, Washington, D.C.
190. National Research Council. (1989). Improving Risk Communication. National Academy Press, Washington, D.C.
191. National Research Council. (1990). Research Strategies for the USGCRP. National Academy Press, Washington, D.C.
192. National Research Council. (1992). Global Environmental Change: Understanding the Human Dimensions, P. C. Stern, O. R. Young, & Druckman, D., (eds.) National Academy Press, Washington, D.C.
193. National Research Council. (1993). Population and Land Use in Developing Countries, C. L. Jolly, & B, B. Torrey, (eds.) National Academy Press, Washington, D.C.
194. National Research Council. (1994a). Assigning Economic Value to Natural Resources. National Academy Press, Washington, D.C.
195. National Research Council. (1994b). Science and Judgment in Risk Assessment. National Academy Press, Washington, D.C.
196. National Research Council. (1994c). Science Priorities for the Human Dimensions of Global Change. National Academy Press, Washington, D.C.
197. National Research Council. (1996a). Learning to Predict Climate Variations Associated with El Niño and the Southern Oscillation. National Academy Press, Washington, D.C.
198. National Research Council. (1996b). Understanding Risk: Informing Decisions in a Democratic Society, P. C. Stern, & V, H. Fineberg, (eds.) National Academy Press, Washington, D.C.
199. National Research Council. (1996c). Upstream: Salmon and Society

in the Pacific Northwest. National Academy Press, Washington, D.C.

200. National Research Council. (1997). Environmentally Significant Consumption: Research Issues, P. C. Stern et al., (eds.) National Academy Press, Washington, D.C.
201. National Research Council. (1998). People and Pixels: Linking Remote Sensing and Social Science. D. Liverman et al., (eds.) National Academy Press, Washington, D.C.
202. National Research Council. (1999). Making Climate Forecasts Matter. P. C. Stern, & E, W. Easterling, (eds.) National Academy Press, Washington, D.C.
203. Nelson, R. R., & Winter, S. G. (1982). An Evolutionary Theory of Economic Change. Harvard University Press, Cambridge, Mass.
204. Netting, R. (1981). Balancing on an Alp. Cambridge University Press, Cambridge, U.K.
205. Nichols, A. L. (1984). Targeting Economic Incentives for Environmental Protection. MIT Press, Cambridge, Mass.
206. Nordhaus, W. D. (1991). To slow or not to slow: The economics of the greenhouse effect. The Economic Journal 101(407), 920–937.
207. Nordhaus, W. D. (1994). Managing the Global Commons: The Economics of Climate Change. MIT Press, Cambridge, Mass.
208. Nordhaus, W. D., & Yohe, G. W. (1983). Future carbon dioxide emissions from fossil fuels. Chapter 2 in Changing Climate. National Academy Press, Washington, D.C.
209. North, D.C. (1990). Institutions, Institutional Change, & Economic Performance. Cambridge University Press, Cambridge, U.K.
210. North, D.C. (1994). Constraints on institutional innovation: Transaction costs, incentive compatibility, and historical considerations. pp. 48–70 in Agriculture, Environment and Health: Sustainable Development in the 21st Century, V. W. Ruttan, ed. University of Minnesota Press, Minneapolis.
211. Ojima, D. S., Parton, W. J., & Schimel, D. S. (1993). Modeling the effects of climatic and CO2 changes on grassland storage of soil, C. Water, Air, & Soil Pollution *70,* 643.
212. Ortloff, C., & Kolata, A. L. (1993). Climate and collapse: Agroecological perspectives on the decline of the Tiwanaku state. Journal of Archaeological Science *20,* 195–221.

213. Ostrom, E. (1990). Governing the Commons: The Evolution of Institutions for Collective Action. Cambridge University Press, New York.

214. Ostrom, E., Gardner, R., & Walker, J. (1994). Rules, Games, & Common-Pool Resources. University of Michigan Press, Ann Arbor.

215. Parry, M. L., Carter, T. R., & T, N. Konijn, (eds.) (1988). The Impact of Climatic Variations on Agriculture. Kluwer Academic Publishers, Boston.

216. Parson, E. A. (1993). Protecting the ozone layer: The evolution and impact of international institutions. pp. 27–73 in Institutions for the Earth: Sources of Effective International Environmental Protection, P. M. Haas, R. O. Keohane, & A, M. Levy, (eds.) MIT Press, Cambridge, Mass.

217. Parson, E. A. (1996). A global climate-change policy exercise: Results of a test run. Working Paper WP-96–90. International Institute for Applied Systems Analysis, Laxenberg, Austria.

218. Parson, E. A. (1997). Informing global environmental policy-making: A plea for new methods of assessment and synthesis. Environmental Modeling and Assessment 2(4).

219. Parson, E. A., & Fisher-Vanden, K. (1997). Integrated assessment models of global climate change. Annual Review of Energy and the Environment *22,* 589–628.

220. Parton, W. J., Schimel, D., & Ojima, D. S. (1994). Environmental change in grasslands: Assessment using models. Climatic Change *28,* 111.

221. Patz, J. A., Epstein, P. R., Burke, T. A., & Balbus, J. M. (1996). Global climate change and emerging infectious diseases. Journal of the American Medical Association *275,* 217–223.

222. Peck, S. C. (1993). CO_2 emissions control comparing policy instruments. Energy Policy 21(March):222–230.

223. Peck, S. C., & Teisberg, T. J. (1992). CETA: A model for carbon emissions trajectory assessment. The Energy Journal *13*(1)*,* 55–77.

224. Pielke, R. A., Jr., & Landsea, C. W. (1998). Normalized hurricane damages in the United States: 1925–1995. Weather and Forecasting *13,* 621–631.

225. Pielke, R. A., Jr., & A, R. Pielke, Sr. (1997). Vulnerability to hurricanes along the, U.S. Atlantic and Gulf coasts: Considerations of the use

of long-term forecasts. pp. 147–184 in Hurricanes: Climate and Socioeconomic Impacts, H. F. Diaz, & S, R. Pulwarty, (eds.) Springer, Berlin.

226. Portney, P. R. (1994). The contingent valuation debate: Why economists should care. Journal of Economic Perspectives *8,* 3–17.
227. Powell, W. W., & DiMaggio, P., (eds.) (1991). The New Institutionalism in Organizational Analysis. University of Chicago Press, Chicago.
228. Pritchett, L. (1994). Desired fertility and the impact of population policies. Population and Development Review *20*(1)*,* 1–55.
229. Ramlogan, R. (1996). Environmental refugees: A review. Environmental Conservation *23*(1)*,* 81.
230. Reilly, J. (1996). Agriculture in a changing climate: Impacts and adaptation. pp. 427–467 in Intergovernmental Panel on Climate Change. Climate Change 1995, Impacts, Adaptation and Mitigation of Climate Change: Scientific and Technical Analyses, R. T. Watson et al., (eds.) Cambridge University Press, Cambridge, U.K.
231. Renard, Y. (1991). Institutional challenges for community-based management in the Caribbean. Nature and Resources *27*(4)*,* 4–9.
232. Renn, O., Webler, T., Rakel, H., Dienel, P., & Johnson, B. (1993). Public participation in decision making: A three-step procedure. Policy Sciences *26,* 189–214.
233. Renn, O., Webler, T., & Wiedemann, P. (1995). Fairness and Competence in Citizen Participation. Kluwer Academic Publishers, Dordrecht, Netherlands .
234. Richards, J. F. (1990). Land transformation. pp. 163–178 in The Earth as Transformed by Human Action, B. L. Turner et al., (eds.) Cambridge University Press, New York.
235. Richardson, B. C. (1992). The Caribbean in the Wider World, 1492–1*992,* A Regional Geography. Cambridge University Press, New York.
236. Richels, R. G., & Edmonds, J. (1995). The economics of stabilizing atmospheric CO2 concentrations. Energy Policy *23,* 373–378.
237. Riebsame, W. E. (1990). The United States Great Plains. pp. 561–575 in The Earth as Transformed by Human Action: Global and Regional Changes in the Biosphere Over the Past 300 Years, B. L. Turner et al., (eds.) Cambridge University Press, Cambridge, U.K.
238. Rokeach, M. (1973). The Nature of Human Values. Free Press, New York.

239. Rosenberg, N. J., Crosson, P. R., K. D. Frederick, W. E. Easterling III, & McKenney, M. S. (1993). The MINK methodology: Background and baseline (Missouri, Iowa, Nebraska, Kansas). Climatic Change 24(1–2):7–22.

240. Rosenzweig, C. (1985). Potential CO_2 induced climate effects on North American wheat producing regions. Climatic Change *7,* 367–389.

241. Rosenzweig, C., & Parry, M. (1994). Potential impact of climate change on world food supply. Nature *367,* 133–138.

242. Rotmans, J. (1990). IMAGE: An Integrated Model to Assess the Greenhouse Effect. Kluwer Academic Publishers, Boston.

243. Rudel, T. K. (1989). Population, development, and tropical deforestration: A cross-national study. Rural Sociology *54,* 327–338

244. Runge, C. F. (1995). Trade, pollution, and environmental protection. pp. 353–375 in The Handbook of Environmental Economics, D. W. Bromley, ed. Blackwell, Oxford, U.K.

245. Runge, C. F., F. Ortalo-Magne, & P. Vande Kamp. (1994). Free Trade, Protected Environment: Balancing Trade Liberalization and Environmental Interests. Council on Foreign Relations, New York.

246. Ruttan, V. W. (1997). Induced innovation, evolutionary theory, and path dependence: Sources of technical change. Economic Journal *107,* 1520–1529.

247. Sagoff, M. (1998). Aggregation and deliberation in valuing environmental public goods: A look beyond contingent pricing. Ecological Economics *24,* 213–230.

248. Sauer, C. O. (1963). Land and Life. University of California Press, Berkeley.

249. Schipper, L., & Martinot, E. (1993). Decline and rebirth: Energy demand in the former USSR. Energy Policy *21*(9), 969–975.

250. Schipper, L., & Meyers, S. (1993). Using scenarios to explore future energy demand in industrialized countries. Energy Policy *21*(3), 264–275.

251. Schipper, L., Meyers, S., Howarth, R., & Steiner, R. (1992). Energy Efficiency and Human Activity: Past Trends, Future Prospects. Cambridge University Press, Cambridge, U.K.

252. Schlager, E., & Ostrom, E. (1992). Property-rights regimes and natural resources: A conceptual analysis. Land Economics *68*(3), 249–262.

253. Schlesinger, W. H., ed. (1997). Biogeochemistry: An Analysis of Global Change. Academic Press, San Diego.

254. Schurr, S. H. (1984). Energy use, technological change, and production efficiency: An economic-historical interpretation. Annual Review of Energy *9,* 409–425.

255. Scott, F. R. (1995). Institutions and Organizations. Sage Publications, Thousand Oaks, Calif.

256. Scott, M. J. (1996). Human settlements in a changing climate: Impacts and adaptation. pp. 399–426 in Intergovernmental Panel on Climate Change. Climate Change 1*995,* Impacts, Adaptation and Mitigation of Climate Change: Scientific and Technical Analyses, R. T. Watson et al., (eds.) Cambridge University Press, Cambridge, U.K.

257. Schwartz, S. H. (1992). Universals in the content and structure of values: Theoretical advances and empirical tests in 20 countries. Advances in Experimental Social Psychology *25,* 1–65.

258. Shafik, N. (1994). Economic development and environmental quality: An econometric analysis. Oxford Economic Papers *46,* 757–773.

259. Shavell, S. (1985). Economic Analysis of Accident Law. Harvard University Press, Cambridge, Mass.

260. Shimata, I., Schaaf, C. B., Thompson, L. G., & E. Moseley-Thompson. (1991). Cultural impacts of severe drought in the prehistoric Andes. World Archaeology *22,* 247–270.

261. Sinclair, P. R. (1987). State Intervention and the Newfoundland Fisheries: Essays on Fisheries Policy and Social Structure. Avebury, Aldershot.

262. Sklair, L. (1991). Sociology of the Global System. Harvester Wheatsheaf, London.

263. Skole, D. L., Chomentowski, W. H., & Salas, W. A. (1994). Physical and human dimensions of deforestation in Amazonia. BioScience *44*(5), 314–322.

264. Slovic, P. (1987). Perception of risk. *Science 236,* 280–285.

265. Smith, C. L. (1986). The life cycle of fisheries. Fisheries *11*(4), 20–24.

266. Social Learning Group. (1998). Social Learning in the Management of Global Environmental Risks, W. C. Clark, J. Jaeger, & J. van Eijndhoven, (eds.) MIT Press, Cambridge, Mass.

267. Socolow, R., Andrews, C., Berkout, F., & Thomas, V., (eds.) (1994).

Industrial Ecology and Global Change. Cambridge University Press, Cambridge, U.K.

268. Solow, R. (1993). An almost practical step toward sustainability. Resources Policy (Sept.): 162–172.
269. Squires, D., Kirkley, J., & Tisdell, C. A. (1995). Individual transferable quotas as a fisheries management tool. Reviews in Fisheries Science *3*(2), 141–169.
270. Steele, J. H. (1996). Regime shifts in fisheries management. Fisheries Research *25,* 19–23.
271. Stern, P. C. (1992). What psychology knows about energy conservation. American Psychologist *47,* 1224–1232.
272. Stern, P. C. (1997). Toward a working definition of consumption for environmental research and policy. pp. 12–25 in Environmentally Significant Consumption: Research Directions. P. C. Stern et al., (eds.) National Academy Press, Washington, D.C.
273. Stern, P. C., & Dietz, T. (1994). The value basis of environmental concern. Journal of Social Issues *50*(3), 65–84.
274. Stern, P. C., & Oskamp, S. (1987). Managing scarce environmental resources. pp. 1043–1088 in Handbook of Environmental Psychology, D. Stokols, & Altman, I., (eds.) Wiley, New York.
275. Stern, P. C., Aronson, E., J. M. Darley, Hill, D., Hirst, E., W. Kempton, & Wilbanks, T. (1986). The effectiveness of incentives for residential energy conservation. Evaluation Review *10,* 147–176.
276. Stern, P. C., Dietz, T., & Kalof, L. (1993). Value orientations, gender, and environmental concern. Environment and Behavior *25*(3), 322–348.
277. Stonich, S., Bort, J., & Ovares, L. (1997). Globalization of the shrimp mariculture industry: The social justice and environmental quality implications in Central America. Society and Natural Resources *10*(2), 161–179.
278. Steward, J. H. (1955). Theory of Cultural Change. University of Illinois Press, Urbana.
279. Thacher, P. S. (1993). The Mediterranean: A new approach to marine pollution. pp. 110–134 in International Environmental Negotiation, G. Sjostedt, ed. Sage, Newbury Park, Calif.
280. Thomas, W. L. (1954). Man's Role in Changing the Face of the Earth. University of Chicago Press, Chicago.

281. Tietenberg, T. H. (1985). Emissions Trading: An Exercise in Reforming Pollution Policy. Resources for the Future, Washington, D.C.

282. Tietenberg, T. H. (1991). Managing the transition: The potential role for economic policies. pp. 187–226 in Preserving the Global Environment. The Challenge of Shared Leadership, J. T. Mathews, ed. Norton, New York.

283. Tietenberg, T. H. (1992). Environmental and Natural Resource Economics, 3d ed. Harper Collins, New York.

284. Tol, R. S. J. (1995). The damage costs of climate change: Toward more comprehensive calculations. Environmental and Resource Economics *5,* 353–374.

285. Toth, F. L. (1994). Models and games for long-term policy problems. Unpublished paper presented at the 1994 meeting of the International Simulation and Gaming Association, Ann Arbor, Mich.

286. Tucker, R. P., & F, J. Richards, (eds.) (1983). Global Deforestation and the Nineteenth-Century World Economy. Duke University Press, Durham, N. C.

287. Turner, B. L., II. et al. (1990). The Earth as Transformed by Human Action: Global and Regional Changes in the Biosphere Over the Past 300 Years. Cambridge University Press, Cambridge, U.K.

288. Turner, B. L., II, G. Hayden, & Kates, R. W. (1993). Population Growth and Agricultural Change in Africa. University Press of Florida, Gainesville.

289. Turner, B. L., II, W. B. Meyer, & Skole, D. L. (1994). Global land-use/land-cover change: Towards an integrated study. Ambio *23*(1)*,* 91.

290. Turner, B. L., II, Skole, D., Sanderson, S., Fischer, G., L. Fresco, & Leemans, R. (1995). Land Use and Land Cover Change: Science/Research Plan. IGBP Rcport No. 35/IIDP Rcport No. 7. Human Dimensions of Global Environmental Change Programme, Geneva.

291. U.S. Department of Energy. (1997). Online International Energy Databases. http://www.eia.doe.gov/fueloverview.html1#international.

292. U.S. Environmental Protection Agency. (1989). Policy Options for Stabilizing Global Climate, D. A. Lashof, & A, D. Tirpak, (eds.) Draft Report to Congress. U.S. Environmental Protection Agency, Office of Policy, Planning, & Evaluation, Washington, D.C.

293. U.S. Global Change Research Program. (1997). Our Changing Planet. USGCRP, Washington, D.C. van Asselt, M. B. A., A. H. W. Beusen,

& M, H. B. Hilderink. (1996). Uncertainty in integrated assessment: A social scientific perspective. Environmental Modeling and Assessment 1(June):71–90.

294. Vari, A., & Tamas, P., (eds.) (1993). Environment and Democratic Transition: Policy and Politics in Central and Eastern Europe. Kluwer Academic Publishers, Boston.
295. Vaughan, E. (1993). Individual and cultural differences in adaptation to environmental risks. American Psychologist *48,* 673–680.
296. Vaughan, E. (1995). The significance of socioeconomic and ethnic diversity for the risk communication process. Risk Analysis *15,* 169–180.
297. Vickers, D., ed. (1997). Marine resources and human societies in the North Atlantic since (1500). Paper presented at the Conference on Marine Resources and Human Societies in the North Atlantic Since 1500, Oct. 20–22, (1995). ISER Conference Paper Number 5. Institute of Social and Economic Research, Memorial University of Newfoundland, St. John's, Newfoundland, Canada.
298. Victor, D. G., Raustiala, K., & B, E. Skolnikoff, (eds.) (1997). The Implementation and Effectiveness of International Environmental Commitments. MIT Press, Cambridge, Mass
299. Viscusi, W. K., & Moore, M. J. (1989). Rates of time preference and valuations of life. Journal of Public Economics *38,* 297–317.
300. von Winterfeldt, D., & Edwards, W. (1986). Decision Analysis and Behavioral Research. Cambridge University Press, New York.
301. Watts, M. J., & Bohle, H. G. (1993). The space of vulnerability: The causal structure of hunger. Progress in Human Geography *17,* 43.
302. Wenger, D. (1985). Collective behavior and disaster research. In Disasters: Theory and Research, E. L. Quarantelli, ed. Sage Publications, Beverly Hills, Calif.
303. Wernick, I. (1996). Consuming materials: The American way. Technological Forecasting and Social Change *53,* 111–122.
304. Wernick, I., & Ausubel, J. (1995). National materials flows and the environment. Annual Review of Energy and the Environment *20,* 462–492.
305. Weyant, J., Davidson, O., Dowlatabadi, H., J. Grubb Edmonds, E. A. Parson, Richels, R., Rotmans, J., P. R. Shukla, R. S. J. Tol, W. Cline, & Fankhauser, S. (1996). Integrated assessment of climate change: An

overview and comparison of approaches and results. pp. 367–396 in Climate Change 1*995,* Economic and Social Dimensions of Climate Change. Contribution of Working Group III to the Second Assessment Report of the Intergovernmental Panel on Climate Change, J. P. Bruce, H. Lee, & F, E. Haites, (eds.) Cambridge University Press, Cambridge, U.K.

306. Wigley, T. M. L., Richels, R. G., & Edmonds, J. A. (1996). Economic and environmental choices in the stabilization of atmospheric C02 concentrations. Nature 379(Jan. 18):240–43.
307. Wilk, R. R. (1997). Emulation and global consumerism. pp. 110–115 in Environmentally Significant Consumption: Research Directions, P. C. Stern et al., (eds.) National Academy Press, Washington, D.C.
308. Witherspoon, S., Mohler, P. P., & Harkness, J. A. (1995). Report on Research into Environmental Attitudes and Perceptions (REAP). European Consortium for Comparative Social Surveys.
309. Wood, W. B. (1994). Forced migration: Local conflicts and international dilemmas. Annals of the Association of American Geographers *84*(4), 607.
310. World Resources Institute. (1996). World Resources: A Guide to the Global Environment. Oxford University Press, New York.
311. Worster, D. ed. (1988). The Ends of the Earth: Perspectives on Modern Environmental History. Cambridge University Press, New York.
312. Yohe, G. W., & Wallace, R. (1996). Near term mitigation policy for global change under uncertainty: Minimizing the expected cost of meeting unknown concentration thresholds. Environmental Modeling and Assessment 1(June):47–57.
313. Yohe, G. W., Neumann, J. E., Marshall, P. B., & Ameden, H. (1996). The economic cost of greenhouse induced sea level rise for developed property in the United States. Climatic Change *32,* 387–410.
314. Young, M. D., & McCay, B. J. (1995). Building equity, stewardship, and resilience into market-based property rights systems. pp. 87–102 in Property Rights and the Environment: Social and Ecological Issues , S. Hanna, & Munasinghe, M., (eds.) The World Bank, Washington, D.C.
315. Young, O. R. (1994a). International Governance: Protecting the Environment in a Stateless Society. Cornell University Press, Ithaca, N.Y.

316. Young, O. R. (1994b). The problem of scale in human/environment relations. Journal of Theoretical Politics *6,* 429–447.

317. Young, O. R. (1996). Rights, rules, and responsibilities in international society. pp. 245–264 in Rights to Nature, S. Hanna, ed. Island Press, Washington, D.C.

318. Young, O. R., & Osherenko, G., (eds.) (1993). Polar Politics: Creating International Environmental Regimes. Cornell Studies in Political Economy. Cornell University Press, Ithaca, N.Y.

319. Zeckhauser, R. (1975). Procedures for valuing lives. Public Policy *23,* 427–463.

Glossary

- *Accuracy:* The degree to which a measurement or observation reflects the actual value. Compare with precision.
- *Acid rain:* The wet deposition only of acidifying substances from the atmosphere. See also acidifying deposition.
- *Acid shock:* An event of relatively acidic surface water that can occur in the springtime when the snowpack melts quickly but the ground is still frozen.
- *Acid sulfate soil:* Acidic soil conditions caused when certain wetlands are drained and sulfide compounds become oxidized.
- *Acid-mine drainage:* Acidic water and soil conditions that develop when sulfide minerals become exposed to the atmosphere, allowing them to be oxidized by Thiobacillus bacteria.
- *Acid-neutralizing capacity:* The quantitative ability of water to neutralize inputs of acid without becoming acidified. See also buffering capacity.
- *Acidification:* An increasing concentration of hydrogen ions ($H+$) in soil or water.
- *Acidifying deposition:* Both the wet and dry deposition of acidifying substances from the atmosphere.
- *Acute toxicity:* Toxicity associated with short-term exposures to chemicals in concentrations high enough to cause biochemical or anatomical damages, even death. Compare with chronic toxicity.

- *Aerobic:* Refers to an environment in which oxygen (O2) is readily available. Compare with anaerobic.
- *Aesthetic pollution:* Substantially a matter of cultural values, this commonly involves images that are displeasing to many (but not necessarily all) people.
- *Afforestation:* Establishment of a forest where one did not recently occur, as when trees are planted on agricultural land.
- *Age-class structure:* The proportions of individuals in various age classes of a population.
- *Agricultural site capability:* See site capability.
- *Agroecosystem:* An ecosystem used for the production of food.
- *Agroforestry:* The cultivation of trees in plantations, typically using relatively intensive management practices.
- *Algal bloom:* An event of high phytoplankton biomass.
- *Ammonification:* Oxidation of the organically bound nitrogen of dead biomass into ammonium (NH4+).
- *Anaerobic:* Refers to an environment in which oxygen (O_2) is not readily available. Compare with aerobic.
- *Angiosperm:* Flowering plants that have their ovules enclosed within a specialized membrane and their seeds within a seedcoat. Compare with gymnosperm.
- *Anthropocentric world view:* This considers humans as being more worthy than other species and uniquely disconnected from nature. The importance and worth of everything is considered in terms of the implications for human welfare. Compare with biocentric world view and ecocentric world view.
- *Anthropogenic:* Occurring as a result of a human influence.
- *Applied ecology:* The application of ecological principles to deal with economic and environmental problems.
- *Aquaculture:* The cultivation of fish and other aquatic species.
- *Aquifer:* Groundwater resources in some defined area.
- *Artificial selection:* The deliberate breeding of species to enhance traits that are viewed as desirable by humans.
- *Artificial wetland:* An engineered wetland, usually constructed to treat sewage or other organic wastes.

- *Aspect:* The direction in which a slope faces.
- *Assimilation efficiency:* In an animal, the percentage of the energy content of ingested food that is absorbed across the gut wall. In plants, the percentage of solar visible light that is fixed by photosynthesis. The term may also be used to refer to the percentage assimilation of ingested inorganic nutrients (such as nitrate or phosphate) by plants or animals, or of drugs by animals.
- *Atmosphere:* The gaseous envelope surrounding the Earth, held in place by gravity.
- *Atmospheric inversion (temperature inversion):* A relatively stable atmospheric condition in which cool air is trapped beneath a layer of warmer air.
- *Atmospheric water:* Water occurring in the atmosphere, in vapor, liquid, or solid forms.
- *Atom bomb:* An explosive device that is based on the uncontrolled "splitting" of certain fissile isotopes of uranium and/or plutonium.
- *Autecology:* The field within ecology that deals with the study of individuals and species. Compare with synecology.
- *Autotroph:* An organism that synthesizes its biochemical constituents using simple inorganic compounds and an external source of energy to drive the process. See also primary producer, photoautotroph, and chemoautotroph.
- *Available concentration:* The concentration of metals in an aqueous extract of soil, sediment, or rocks, simulating the amount available for organisms to take up from the environment. compare with total concentration.
- *Baby boom:* A period of high fecundity during 1945–1965 that occurred because of social optimism after the second world war.
- *Background concentration:* A presence or concentration of a substance that is not significantly influenced by either anthropogenic emissions or unusual natural exposures.
- *Binomial:* Two latinized words that are used to name a species.
- *Bioaccumulation (bioconcentration):* The occurrence of chemicals in much higher concentrations in organisms than in the ambient environment. Compare with food-web magnification.

- *Biocentric world view:* This considers all species (and individuals) as having equal intrinsic value. humans are not considered more important or worthy than any other species. compare with anthropocentric world view and ecocentric world view.
- *Bioconcentration:* See Bioaccumulation.
- *Biodegradation:* The breakdown of organic molecules into simpler compounds through the metabolic actions of microorganisms.
- *Biodiversity:* The richness of biological variation, including genetic variability as well as species and community richness.
- *Biodiversity crisis:* THE present era of high rates of extinction and endangerment of biodiversity.
- *Biogeochemical prospecting:* Prospecting for metal ores using observations of high metal concentrations in plants, soil, or surface rocks.
- *Biological control:* Pest-control methods that depend on biological interactions, such as diseases, predators, or herbivores. biological oxygen demand (bod): the capacity of organic matter and other substances in water to consume oxygen during decomposition.
- *Biomagnification:* See food-web magnification.
- *Biomass energy:* The chemical potential energy of plant biomass, which can be combusted to provide thermal energy.
- *Biome:* A geographically extensive ecosystem, occurring throughout the world wherever environmental conditions are suitable.
- *Biophilia:* An innate love of people for nature.
- *Bio-resource:* A renewable resource that is biological in character.
- *Biosphere:* All life on earth, plus their ecosystems and environments.
- *Birth Control:* Methods used to control fertility and childbirth.
- *BOD:* See biological oxygen demand.
- *BOG:* An infertile, acidic, unproductive wetland that develops in cool but wet climates. compare with fen.
- *Boreal coniferous forest:* A northern forest dominated by coniferous trees, usually species of fir, larch, pine, or spruce. see also, boreal forest.

- *Boreal forest (Taiga):* An extensive biome occurring in environments with cold winters, short but warm growing seasons, and moist soils, and usually dominated by coniferous trees.
- *Broad-spectrum pesticide:* A pesticide that is toxic to other organisms as well as the pest.
- *Broadcast spray:* A pesticide treatment over a large area.
- *Browse:* Broad-leaved shrubs that are eaten by herbivores such as hares and deer.
- *Bryophyte:* Simple plants that do not have vascular tissues nor a cuticle on their foliage.
- *By-catch:* Inadvertent harvesting of a non-target species.
- *Buffering capacity:* The ability of a solution to resist changes in ph as acid or base is added.
- *Calorie:* A standard unit of energy, defined as the amount of energy needed to raise the temperature of one gram of pure water from 15°C to 16°C. Compare with joule.
- *Carbon credits:* Actions that help reduce the atmospheric concentration of CO_2, such as fossil-fuel conservation and planting trees.
- *Carnivore (secondary consumer):* An animal that hunts and eats other animals.
- *Carrying capacity:* The abundance of a species that can be sustained without the habitat becoming degraded.
- *Chaparral:* A shrub-dominated ecosystem that occurs in south-temperate environments with winter rains and summer drought.
- *Chemical weapons:* Weapons that cause deaths or injuries through exposure to toxic chemicals.
- *Chemoautotroph:* Microorganisms that harness some of the potential energy of certain inorganic chemicals (e.g., sulfides) to drive their fixation of energy through chemosynthesis. Compare with photoautotroph.
- *Chemosynthesis:* Autotrophic productivity that utilizes energy released during the oxidation of certain inorganic chemicals (such as sulfides) to drive biosynthesis. Compare with photosynthesis.
- *Chromosome:* Subcellular unit composed of DNA and containing the genetic information of eukaryotic organisms.

- *Chronic Toxicity:* Toxicity associated with exposure to small or moderate concentrations of chemicals, sometimes over a long period of time. the damages may be biochemical or anatomical, and may include the development of a lethal disease, such as cancer. Compare with acute toxicity.
- *Clear-cutting:* The harvesting of all economically useful trees from an area at the same time.
- *Climate:* The prevailing, long-term, meteorological conditions of a place or region, including temperature, precipitation, wind speed, and other factors. compare with weather.
- *Climate Change:* Long-term changes in air, soil, or water temperature; precipitation regimes; wind speed; or other climate-related factors.
- *Coal:* An organic-rich, solid fossil fuel mined from sedimentary geological formations.
- *Coal Washing:* See fuel desulfurization.
- *Coarse Woody Debris:* Logs lying on the forest floor.
- *Coevolution:* This occurs when species interact in ways that affect their reciprocal survival, and so are subject to a regime of natural selection that reinforces their mutual evolutionary change.
- *Collective Properties:* This term is used in reference to the summation of the parts of a system. see also, emergent properties.
- *Commensalism:* A symbiosis in which one of the species benefits from the interaction, while the other is not affected in either a positive or negative way.
- *Commercial energy production:* The use of solid, liquid, and gaseous fuels, plus all electricity. does not include the use of traditional fuels. see also, total energy production and traditional fuels.
- *Commercial extinction:* Depletion of a natural resource to below the abundance at which it can be profitably harvested.
- *Common-property Resource:* A resource shared by all of society, not owned by any particular person or interest.
- *Community:* In ecology, this refers to populations of various species that are co-occurring at the same time and place.

- *Community-replacing disturbance:* A disturbance that results in the catastrophic destruction of an original community, and its replacement by another one. Compare with microdisturbance.
- *Compaction:* A decrease in the pore space of soil (or increased bulk density) caused by the passage of heavy machinery.
- *Compartment:* A reservoir of mass in a nutrient or material cycle.
- *Competition:* A biological interaction occurring when the demand for an ecological resource exceeds its limited supply, causing organisms to interfere with each other.
- *Competitor:* A species that is dominant in a habitat in which disturbance is rare and environmental stresses are unimportant, so competition is the major influence on evolution and community organization.
- *Compost:* Partially decomposed, well-humified organic material.
- *Composting:* The processing of discarded organic material by encouraging decomposition processes under warm, moist, oxygen-rich conditions. the product, known as compost, is a useful fertilizer and soil conditioner.
- *Conservation:* Wise use of natural resources. conservation of nonrenewable resources involves recycling and other means of efficient use. conservation of renewable resources includes these means, in addition to ensuring that harvesting does not exceed the rate of regeneration of the stock.
- *Contamination:* The presence of potentially damaging chemicals in the environment, but at concentrations less than those required to cause toxicity or other ecological damages. compare with pollution.
- *Control (control treatment):* An experimental treatment that was not manipulated, and is intended for comparison with manipulated treatments.
- *Conventional economics:* Economics as it is commonly practised, which includes not accounting for costs associated with ecological damages and resource depletion. compare with ecological economics.
- *Conventional munitions:* Explosive devices that are based on chemical reactions, such as cordite and dynamite.

- *Convergence (evolutionary conversion):* This occurs when unrelated species with similar niches and living in comparable environments are subjected to parallel regimes of natural selection, resulting in their evolution to be similar in morphology, physiology, and behavior.
- *Conversion:* See ecological conversion.
- *Core:* Earth's massive interior, made up of hot molten metals.
- *Coriolis effect:* An influence of earth's west-to-east rotation, which makes winds in the northern hemisphere deflect to the right and those in the southern hemisphere to the left.
- *Creationist:* A person who rejects the theory of evolution in favour of a literal interpretation of genesis, the first book of the old testament of the bible. see also, scientific creationist.
- *Critical load:* A threshold for pollutant inputs, below which it is thought ecological damages will not be caused.
- *Crude Oil:* See Petroleum.
- *Crude oil washing (cow) method:* A method of washing a tanker's oil-storage components with a spray of crude oil before the next cargo is loaded. This eliminates the use of wash-water and avoids an important cause of marine oil pollution.
- *Crust:* The outermost layer of earth's sphere, overlying the lithosphere and composed mostly of crystalline rocks.
- *Cultural eutrophication:* Eutrophication caused by anthropogenic nutrient inputs, usually through sewage dumping or fertilizer runoff. See also, eutrophication.
- *Cultural evolution:* Adaptive evolutionary change in human society, characterized by increasing sophistication in the methods, tools, and social organizations used to exploit the environment and other species. compare with evolution.
- *Cultural identity:* A complex of self-identified characteristics and values that a group of people considers important in defining their distinct quality.
- *Culture:* The shared beliefs, values, and knowledge of a defined group of people.
- *Cumulative environmental impacts:* Environmental impacts that result from a proposed undertaking, in addition to those caused by any past, existing, and imminent developments and activities.

- *Decay:* The decomposition or oxidation of dead biomass, mostly through the actions of microorganisms.
- *Decomposer:* See Detritivore.
- *Deductive logic:* Logic in which initial assumptions are made and conclusions are then drawn from those assumptions. compare with inductive logic.
- *Deep drainage:* Soil water that has drained to below the lower limits of plant roots.
- *Deforestation:* A permanent conversion of forest into some other kind of ecosystem, such as agriculture or urbanized land use.
- *Demographic transition:* A change in human population parameters from a condition of high birth and death rates to one of low birth and death rates.
- *Denitrification:* The microbial reduction of nitrate (NO^{3-}) into gaseous N_2O or N_2.
- *Desert:* A temperate or tropical biome characterized by prolonged drought, usually receiving less than 25 cm of precipitation per year.
- *Desertification:* The increasing aridity of drylands; an environmental change that can make agriculture difficult or impossible.
- *Detritivore:* A heterotroph that feeds on dead organic matter.
- *Developed countries:* Countries with a relatively well-organized economic infrastructure and a high average per-capita income. see also, high-income countries and compare with less-developed countries.
- *Development (economic development):* An economic term that implies improving efficiency in the use of materials and energy in an economy, and progress toward a sustainable economic system. compare with economic growth.
- *Discipline:* A specific area of study, such as mathematics or music.
- *Disturbance:* An episode of destruction of some part of a community or ecosystem.
- *DNA:* The biochemical deoxyribonucleic acid, the main constituent of the chromosomes of eukaryotic organisms.

- *Domestication:* The genetic, anatomic, and physiological modification of crops and other species from their wild, progenitor species, through the selective breeding of preferred races (or cultivars).
- *Dose-response relationship:* The quantitative relationship between different doses of a chemical and a biological or ecological response.
- *Doubling time:* The time it takes for something to increase by a factor of two (as in population growth).
- *Drift:* Movement of applied pesticide off the intended site of deposition through atmospheric or aquatic transport.
- *Dry deposition:* Atmospheric inputs of chemicals occurring in intervals between rainfall or snowfall. compare with wet deposition.
- *Dumping:* The long-term disposal of disused material, for example, by placing solid waste into a sanitary landfill, or by discarding liquid waste into a waterbody.
- *Earthquake:* A trembling or movement of the earth, caused by a sudden release of geological stresses at some place within the crust.
- *Ecocentric world view:* This incorporates the biocentric world view but also stresses the importance of interdependent ecological functions, such as productivity and nutrient cycling. in addition, the connections among species within ecosystems are considered to be invaluable. compare with anthropocentric world view and biocentric world view.
- *Ecofeminism:* A philosophical and political movement that applies feminist ideas to environmental concerns.
- *Ecological conversion:* A long-term change in the character of the ecosystem at some place, as when a natural forest is converted into an agricultural land use.
- *Ecological economics:* A type of economics that involves a full accounting of costs associated with ecological damages and resource depletion. compare with conventional economics.
- *Ecological footprint:* The area of ecoscape (i.e., landscape and seascape) required to supply a human population with the

necessary food, materials, energy, waste disposal, and other crucial goods and services.

- *Ecological integrity (ecosystem health):* A notion related to environmental quality, but focusing on changes in natural populations and ecosystems, rather than effects on humans and their economy. see also, environmental quality.
- *Ecological justice:* A worldview in which all species (i.e., not just humans) have a right to equitable access to the necessities of life and happiness. see also, social justice.
- *Ecological pyramid:* A model of the trophic structure of an ecosystem, organized with plant productivity on the bottom, that of herbivores above, and carnivores above the herbivores.
- *Ecological service:* An ecological function that is useful to humans and to ecosystem stability and integrity, such as nutrient cycling, productivity, and control of erosion.
- *Ecological stress:* See stressors.
- *Ecological sustainability:* See ecologically sustainable development.
- *Ecological values:* Broader utilitarian values that are based on the needs of humans, but also on those of other species and natural ecosystems.
- *Ecologically sustainable development:* This considers the human need for resources within an ecological context, and includes the need to sustain all species and all components of earth's life-support system. Compare with sustainable development.
- *Ecologically sustainable economic system:* An economic system that operates without a net consumption of natural resources, and without endangering biodiversity or other ecological values. ultimately, ecologically sustainable economic systems are supported by the wise use of renewable resources.
- *Ecologically sustainable economy:* An economy in which ecological goods and services are utilized in ways that do not compromise their future availability and do not endanger the survival of species or natural ecosystems.
- *Ecology:* The study of the relationships between organisms and their environment.
- *Economic development:* See Development.

- *Economic growth:* A term that refers to an economy that is increasing in size over time, usually due to increases in both population and per capita resource use. compare with development.
- *Ecoregion:* See Ecozone.
- *Ecoscape:* A general term for landscapes or seascapes.
- *Ecosystem:* A general term used to describe one or more communities that are interacting with their environment as a defined unit. ecosystems range from small units occurring in microhabitats, to larger units such as landscapes and seascapes, and even the biosphere.
- *Ecosystem approach:* A holistic interpretation of the natural world that considers the web-like interconnections among the many components of ecosystems.
- *Ecosystem health:* See Ecological integrity.
- *Ecotone:* A zone of transition between two distinct habitats.
- *Ecotoxicology:* Study of the directly poisonous effects of chemicals in ecosystems, plus indirect effects such as changes in habitat or food abundance caused by toxic exposures. compare with toxicology and environmental toxicology.
- *Ecotype:* A population specifically adapted to coping with locally stressful conditions, such as soil with high metal concentrations.
- *Ecozone:* The largest biophysical zones in the national ecological classification of Canada.
- *Electromagnetic energy:* Energy associated with photons, comprising an electromagnetic spectrum divided into components, including ultraviolet, visible, and infrared.
- *Emergent property:* A used in reference to synergetic properties that are greater than the summation of the parts of a system. see also, collective properties.
- *Emissions trading:* A system in which a company whose that has not exceeded its cap on emissions of a regulated substance, such as SO_2, can sell its "surplus" to another that is likely to exceed its cap.
- *Endangered:* In Canada, this specifically refers to indigenous species threatened with imminent extinction or extirpation over all or a significant portion of their canadian range.

- *Endemic:* An ecological term used to describe species with a local geographic distribution.
- *Energy:* The capacity of a body or system to accomplish work, and existing as electromagnetic, kinetic, and potential energies.
- *Energy budget:* An analysis of the rates of input and output of energy to a system, plus transformations of energy among its states, including changes in stored quantities.
- *Energy production:* See total energy production.
- *Entropy:* A physical attribute related to the degree of randomness of the distributions of matter and energy.
- *Environment:* (1) refers to influences on organisms and ecosystems, including both non-living (abiotic) and biological factors; (2) an indeterminate word for issues associated with the causes and consequences of environmental damage, or with the larger environmental crisis.
- *Environmental citizenship:* Actions taken by individuals and families to lessen their impacts on the environment.
- *Environmental degradation:* Refers to pollution, disturbance, resource depletion, lost biodiversity, and other kinds of environmental damage; usually refers to damage occurring accidentally or intentionally as a result of human activities (see also anthropogenic), but can also be caused by natural disasters or stressors.
- *Environmental discrimination (environmental prejudice):* Discrimination against any defined group that results in them suffering a disproportionate amount of degradation or pollution of their living or work environment. See also, environmental racism.
- *Environmental ecology:* See applied ecology.
- *Environmental education:* A way of fostering environmental literacy by incorporating environmental issues in educational curricula, both in specialized classes as well as across the curriculum, and also including the out-of-school public.
- *Environmental ethics:* These deal with the responsibilities of the present human generation to ensure continued access to adequate resources and livelihoods for future generations of people and other species.

- *Environmental impact assessment (EIA):* A process used to identify and evaluate the potential consequences of proposed actions or policies for environmental quality. See also, socioeconomic impact assessment.
- *Environmental indicators:* Relatively simple measurements that are sensitive to changes in the intensity of stressors, and are considered to represent complex aspects of environmental quality.
- *Environmental literacy:* Refers to an objective understanding, by individuals and society-at-large, of the causes and consequences of environmental problems.
- *Environmental monitoring:* Repeated measurements of indicators related to the inorganic environment or to ecosystem structure and function.
- *Environmental mutagen:* A mutagenic influence that is encountered in the environment. see also, mutagen.
- *Environmental non-governmental organizations (ENGOS):* Charities and other not-for-profit organizations that are working in the environmental field. See also, non-governmental organizations.
- *Environmental quality:* A notion related to the amounts of toxic chemicals and other stressors in the environment, to the frequency and intensity of disturbances, and to their effects on humans, other species, ecosystems, and economies.
- *Environmental racism:* Discrimination against a group of people defined by racial attributes, which results in them suffering a disproportionate amount of degradation or pollution of their living or work environment. See also, environmental discrimination.
- *Environmental reporting:* Communication of information about changes in environmental quality to interest groups and the general public.
- *Environmental risk:* A hazard or probability of suffering damage or misfortune because of exposure to some environmental circumstance.
- *Environmental risk assessment:* A quantitative evaluation of the risks associated with an environmental hazard.
- *Environmental science:* An interdisciplinary branch of science that investigates questions related to the human population,

resources, and damages caused by pollution and disturbance.

- *Environmental scientist:* A scientist who is specialized in some aspect of environmental science.
- *Environmental security:* The protection of people and the public interest from environmental risks, particularly those associated with anthropogenic activities and accidents, but may also include natural dangers.
- *Environmental stressor:* See Stressor.
- *Environmental studies:* An extremely interdisciplinary approach that examines the scientific, social, and cultural aspects of environmental issues.
- *Environmental teratogen:* A teratogenic influence that is encountered in the environment. see also, teratogen.
- *Environmental toxicology:* The study of environmental factors influencing exposures of organisms to potentially toxic levels of chemicals. compare with toxicology and ecotoxicology.
- *Environmental values:* Perceptions of the worth of environmental components, divided into two broad classes: utilitarian and intrinsic.
- *Environmentalist:* Anyone with a significant involvement with environmental issues, usually in an advocacy sense.
- *Erosion:* The physical removal of rocks and soil through the combined actions of flowing water, wind, ice, and gravity.
- *Estuary:* A coastal, semi-enclosed ecosystem that is open to the sea and has habitats transitional between marine and freshwater conditions.
- *Ethics:* The perception of right and wrong. the proper behavior of people toward each other and toward other species and nature.
- *Eukaryote:* Organisms in which the cells have an organized, membrane-bound nucleus containing the genetic material. compare with prokaryote.
- *Eutrophic:* Pertains to waters that are highly productive because they contain a rich supply of nutrients. compare with oligotrophic and mesotrophic.
- *Eutrophication:* Increased primary productivity of an aquatic ecosystem, resulting from nutrient inputs.

- *Evaporation:* The change of state of water from a liquid or solid to a gas.
- *Evapotranspiration:* Evaporation of water from a landscape. see also, transpiration.
- *Evolution:* Genetically based changes in populations of organisms, occurring over successive generations.
- *Evolutionary ecology:* The interpretation of ecological knowledge in terms of evolution, natural selection, and related themes.
- *Experiment:* A controlled test or investigation designed to provide evidence for, or preferably against, a hypothesis about the natural or physical world.
- *Exposure:* In ecotoxicology, this refers to the interaction of organisms with an environmental stressor at a particular place and time.
- *Exposure assessment:* An investigation of the means by which organisms may encounter a potentially toxic level of a chemical or other environmental stressor.
- *Externality:* A cost or benefit that is received, even though the affected party did not choose to incur it.
- *Extant:* A species that still exists. compare with extinct.
- *Extinct (Extinction):* A condition in which a species or other taxon no longer occurs anywhere on earth.
- *Extinction crisis:* See Biodiversity crisis.
- *Extinction Vortex:* An accelerating spiral of endangerment and extinction caused by worsening environmental conditions.
- *Extirpated (extirpation):* A condition in which a species or other taxon no longer occurs in some place or region, but still survives elsewhere.
- *Fact:* An event or thing known to have happened, to exist, or to be true. See also, hypothesis.
- *Fen:* A wetland that develops in cool and wet climates, but is less acidic and more productive than a bog because it has a better nutrient supply. compare with bog.
- *First law of thermodynamics:* A physical principle stating that energy can undergo transformations among its various states, but it is never created or destroyed; thus, the energy content

of the universe remains constant. see also, second law of thermodynamics.

- *First nations:* The aboriginal people(s) originally living in some place. this term is often used in reference to the original inhabitants of the americas, prior to the colonization of those regions by europeans, and their modern descendants.
- *Fission bomb:* See atom bomb.
- *Fission reaction:* nuclear Reaction involving the splitting of heavier, radioactive atoms into lighter ones, with the release of large quantities of energy.
- *Fitness:* The proportional contribution of an individual to the progeny of its population.
- *Flow-through system:* a system with an input and an output of energy or mass, plus temporary storage of any difference.
- *Flue-gas desulfurization:* A process to remove SO_2 from the waste (flue) gases of a power plant or smelter, before they are discharged into the atmosphere.
- *Flux:* A movement of mass or energy between compartments of a material or energy cycle.
- *Food chain:* A hierarchical model of feeding relationships among species in an ecosystem.
- *Food web:* A complex model of feeding relationships, describing the connections among all food chains within an ecosystem.
- *Food-web magnification (food-web accumulation, food-web concentration):* The tendency for top predators in a food web to have the highest residues of certain chemicals, especially organochlorines. compare with bioaccumulation.
- *Forest floor:* Litter and other organic debris lying on top of the mineral soil of a forest.
- *Forestry:* The harvesting of trees and management of post-harvest succession to foster the regeneration of another forest.
- *Fossil fuel:* Organic-rich geological materials, such as coal, petroleum, and natural gas.
- *Frontier world view:* This asserts that humans have a right to exploit nature by consuming natural resources in boundless

quantities. see also, sustainability world view and spaceship world view.

- *Fuel desulfurization:* A process that removes much of the sulfur content of coal before it is used as a fuel in a power plant.
- *Fuel switching:* The replacement of a high-sulfur fuel, such as coal, by an energy source that does not emit sulfur gases, such as hydroelectricity or nuclear power.
- *Full-cost accounting system:* An accounting system that considers all costs, including those of environmental damage.
- *Fungicide:* A pesticide used to protect crop plants and animals from fungi that cause diseases or other damages.
- *Fusion bomb:* See hydrogen bomb.
- *Fusion reaction:* Nuclear reaction involving the combining of light nuclei, such as those of hydrogen, to make heavier ones, with the release of large quantities of energy. Fusion reactions occur under conditions of intense temperature and pressure, such as within stars and in hydrogen bombs.
- *Gaia hypothesis:* A notion that envisions earth's species and ecosystems as a "superorganism" that attempts to optimize environmental conditions toward enhancing its own health and survival.
- *Gaseous wastes:* The gaseous products of combustion or industrial reactions.
- *Gene:* A region of a chromosome, containing a length of DNA that behaves as a particulate unit in inheritance and determines the development of a specific trait.
- *Genocide:* The mass killing of an identifiable group as an attempted extermination.
- *Genotype:* The genetic complement of an individual organism. See also, phenotype.
- *Geography:* The study of the features of the surface of the earth, including topography, landforms, soil, climate, and vegetation, as well as the intersections of these with the economic interests of humans.
- *Geothermal energy:* Heat in earth's crust, which can sometimes be used to provide energy for heating or generation of electricity.
- *Glaciation:* An extensive environmental change associated with

an extended period of global climatic cooling and characterized by advancing ice sheets.

- *Glacier:* A persistent sheet of ice, occurring in the arctic and antarctic and at high altitude on mountains.
- *Greater protected area:* A protected area plus its immediately surrounding area, co-managed to sustain populations of indigenous species and natural communities.
- *Green manure:* Living plant biomass that is grown and then incorporated into the soil by tillage.
- *Green revolution:* Intensive agricultural systems involving the cultivation of improved crop varieties in monoculture, and increased use of mechanization, fertilizers, and pesticides.
- *Greenhouse effect:* The physical process by which infrared-absorbing gases (such as CO_2) in Earth's atmosphere help to keep the planet warm.
- *Greenhouses gases (GHGS):* Atmospheric gases that efficiently absorb infrared radiation and then dissipate some of the thermal energy gain by re-radiation. synonym: radiatively active gases.
- *Gross domestic product (GDP):* The total annual value of all goods and services produced domestically within a country. GDP is equivalent to gross national product minus net investment income from foreign countries. See also, gross national product (GNP).
- *Gross national product (GNP):* The total annual value of all goods and services produced domestically by a country, including net foreign investment income. See also, gross domestic product (GDP).
- *Gross primary production (GPP):* The fixation of energy by primary producers within an ecosystem. see also, respiration, net primary production, and autotroph.
- *Groundwater:* Water stored underground in soil and rocks.
- *Groundwater drainage:* The drainage of water to storage places in the ground, occurring under the influence of gravity.
- *Growth:* Refers to an economy or economic sector that is increasing in size over time. Compare with development.
- *Gymnosperm:* Vascular plants such as conifers, which have naked ovules not enclosed within a specialized membrane, and seeds

without a seedcoat. Compare with angiosperm.

- *Habitat:* The place or "home" where a plant or animal lives, including the specific environmental factors required for its survival.
- *Harvesting effort:* The amount of harvesting, which is a function of both the means (such as the kinds of fishing gear) and the intensity (the number of boats and the amount of time each spends fishing).
- *Harvesting mortality:* Anthropogenic mortality, especially that due to the harvesting of a bio-resource. Compare with natural mortality.
- *Hazardous waste:* Wastes that are flammable, explosive, toxic, or otherwise dangerous. see also, toxic waste.
- *Herbicide:* A pesticide used to kill weeds. See also, weed.
- *Herbivore (Or Primary Consumer):* An animal that feeds on plants.
- *Heterotroph:* An organism that utilizes living or dead biomass as food.
- *Hidden injury:* A reduction in plant productivity caused by exposure to pollutants, but not accompanied by symptoms of acute tissue damages.
- *High-income countries:* Countries with a relatively high average per-capita income. See also, developed countries and compare with low-income countries.
- *Hormone:* A biochemical produced in an endocrine gland (and transported by the blood) that functions to regulate a metabolic process. Some chemicals in food may mimic the function of hormones produced naturally in the body.
- *Hormonally active substance:* A hormone or another chemical that has an effect on the regulation of biochemistry. See also, hormone.
- *Humidity:* the actual concentration of water in the atmosphere, usually measured in mg/m^3. compare with relative humidity.
- *Humus:* Amorphous, partially decomposed organic matter. an important and persistent type of soil organic matter. it is very important in soil tilth and fertility.

- *Hydrocarbons:* Molecules composed of hydrogen and carbon only.
- *Hydroelectric energy:* Electricity generated using the kinetic energy.
- *Hydrogen bomb:* A nuclear weapon that is based on the fusion of nuclei of deuterium and tritium, two isotopes of hydrogen.
- *Hydrologic (water) cycle:* The movement between, and storage of water in, various compartments of the hydrosphere. See also, hydrosphere.
- *Hydrosphere:* The parts of the planet that contain water, including the oceans, atmosphere, on land, in surface waterbodies, underground, and in organisms.
- *Hyperaccumulator:* A species that bioaccumulates metals or other chemicals to extremely high concentrations in their tissues. See also, bioaccumulation.
- *Hypereutrophic:* Extremely eutrophic waters; usually considered to be a degraded ecological condition. See also, eutrophic.
- *Hypersensitivity:* An extreme sensitivity to exposure to some environmental factor, resulting in a biological response such as asthma, disease, or even death. It may be expressed at the species or individual level, and it involves responses at relatively low intensities of exposure that the great majority of species or individuals could tolerate.
- *Hypothesis:* A proposed explanation for the occurrence or causes of natural phenomena. scientists formulate hypotheses as statements, and test them through experiments and other forms of research. See also, fact.
- *Igneous rock:* Rock such as basalt and granite, formed by cooling of molten magma.
- *Impoundment:* An area of formerly terrestrial landscape that is flooded behind a dam.
- *Incineration:* The combustion of mixed solid wastes to reduce the amount of organic material present.
- *Indicator:* See environmental indicator.
- *Indigenous culture:* A human culture existing in a place or region prior to its invasion, or other significant influence, by a foreign culture.

- *Individual organism:* A genetically and physically discrete living entity.
- *Inductive logic:* Logic in which conclusions are objectively developed from the accumulating evidence of experience and the results of experiments. see also, deductive logic.
- *Inequitable:* Not equitable or fair.
- *Inherent value:* See intrinsic value
- *Inhumane:* Reflecting a lack of pity or compassion; most commonly refers to the cruel treatment by humans of other animals.
- *Insecticide:* A pesticide used to kill insects that are considered pests. see also, pesticide and pest.
- *Instrumental value:* See utilitarian value.
- *Integrated forest management:* Forest management plans that accommodate the need to harvest timber from landscapes, while also sustaining other values, such as hunted wildlife, outdoor recreation, and biodiversity.
- *Integrated pest management (IPM):* The use of a variety of complementary tactics toward pest control, with the aim of having fewer environmental and health risks.
- *Interdisciplinary:* Encompassing a wide diversity of kinds of knowledge.
- *Intrinsic population change:* Population change due only to the balance of birth and death rates.
- *Intrinsic value:* Value that exists regardless of any direct or indirect value in terms of the needs or welfare of humans.
- *Invasive alien:* Refers to non-native species that survive in wild habitats and possibly aggressively out-compete native species or cause other kinds of ecological damage.
- *Inversion:* See atmospheric inversion.
- *Invertebrate:* Any animal that lacks an internal skeleton, and in particular a backbone.
- *Joule:* A standard unit of energy, defined as the energy needed to accelerate 1 kg of mass at 1 m/s^2 for a distance of 1 meter. compare with calorie.

- *K-selected:* Refers to organisms that produce relatively small numbers of large offspring. a great deal of parental investment is made in each progeny, which helps to ensure their establishment and survival. Compare with R-selected.
- *Keystone species:* A dominant species in a community, usually a predator, with an influence on structure and function that is highly disproportionate to its biomass.
- *Kinetic energy:* Energy associated with motion, including mechanical and thermal types.
- *Knowledge:* information and understanding about the natural world.
- *landscape:* the spatial integration of ecological communities over a large terrestrial area.
- *landscape ecology:* study of the spatial characteristics and temporal dynamics of communities over large areas of land (landscapes) or water (seascapes).
- *laws of thermodynamics:* physical principles that govern all transformations of energy. see also, first law of thermodynamics and second law of thermodynamics.
- *leaching:* the movement of dissolved substances through the soil with percolating rainwater.
- *legacy munitions:* see unexploded ordinance.
- *lentic ecosystem:* a freshwater ecosystem characterized by nonflowing water, such as a pond or lake. compare with lotic ecosystem.
- *less-developed countries:* countries with a relatively well-organized economic infrastructure and a high average per-capita income. see also, high-income countries and compare with developed countries.
- *life form:* a grouping of organisms on the basis of their common morphological and physiological characteristics, regardless of their evolutionary relatedness.
- *life index (production life):* the known reserves of a resource divided by its current rate of production.
- *liming:* treatment of a waterbody or soil to reduce acidity, usually by adding calcium carbonate or calcium hydroxide.

- *limiting factor:* an environmental factor that is the primary restriction on the productivity of autotrophs in an ecosystem. see also, principle of limiting factors.
- *liquid waste:* variable urban wastes that include sewage and discarded industrial and household fluids.
- *lithification:* a geological process in which materials are aggregated, densified, and cemented into new sedimentary rocks.
- *lithosphere:* an approximately 80-km thick region of rigid, relatively light rocks that surround earth's plastic mantle.
- *load-on-top (LOT) method:* a process used in ocean-going petroleum tankers to separate and contain most oily residues before ballast waters are discharged to the marine environment.
- *Lotic ecosystem:* a freshwater ecosystem characterized by flowing water, such as a stream or river. compare with lentic ecosystem.
- *low-income counties:* countries with a relatively small average per-capita income. see also, less-developed countries and compare with high-income countries.
- *LRTAP:* The long-range transport of atmospheric pollutants.
- *Macroclimate:* climatic conditions affecting an extensive area. compare with microclimate.
- *macroevolution:* the evolution of species or higher taxonomic groups, such as genera, families, or classes. compare with microevolution.
- *management system:* a variety of management practices used in a coordinated manner.
- *manipulative experiment:* an experiment involving controlled alterations of factors hypothesized to influence phenomena, conducted to investigate whether predicted responses will occur, thereby uncovering causal relationships. see also, experiment and natural experiment.
- *mantle:* a less-dense region that encloses earth's core, and composed of minerals in a hot, plastic state known as magma.
- *marsh:* a productive wetland, typically dominated by species of monocotyledonous angiosperm plants that grow as tall as several metres above the water surface.

- *mass extinction:* an event of synchronous extinction of many species, occurring over a relatively short period of time. may be caused by natural or anthropogenic forces.
- *maximum sustainable yield (MSY):* the largest amount of harvesting that can occur without degrading the productivity of the stock.
- *mechanization:* the use of specialized machinery to perform work, instead of the labor of people or animals.
- *megacity:* a large city, sometimes defined as having a population greater than 8 million people.
- *mesosphere:* the layer of the atmosphere extending beyond the stratosphere to about 75 km above the surface of the earth. see also, stratosphere.
- *mesotrophic:* pertains to aquatic ecosystems of moderate productivity, intermediate to eutrophic and oligotrophic waters. compare with eutrophic and oligotrophic.
- *metal:* any relatively heavy element that in its pure state shares electrons among atoms, and has useful properties such as malleability, high conductivity of electricity and heat, and tensile strength.
- *metamorphic rock:* rock formed from igneous or sedimentary rocks that have changed in structure under the influences of geological heat and pressure.
- *meteorite:* an extraterrestrial rock-like object; very rarely, one may intersect with earth's orbit and impact the planet.
- *microclimate:* climatic conditions on a local scale. compare with macroclimate.
- *Microdisturbance:* local disruptions that affect small areas within an otherwise intact community. compare with community-replacing disturbance.
- *microevolution:* relatively subtle evolutionary changes occurring within a population or species, sometimes within only a few generations, and at most leading to the evolution of races, varieties, or subspecies. compare with macroevolution.
- *middle-income countries:* countries with a rapidly increasing average per-capita income. see also, high-income and low-

income countries and compare with developed countries and less-developed countries.

- *militarism:* a belief of people or governments in the need to maintain a strong military capability to defend or promote national interests.
- *mitigation:* an action that repairs or offsets environmental damages to some degree.
- *monoculture:* the cultivation of only one species while attempting to exclude others from the agroecosystem.
- *montane forest:* a conifer-dominated forest occurring below the alpine zone on mountains.
- *MSY:* see maximum sustainable yield.
- *mutagen:* a chemical or physical agent (e.g., ultraviolet radiation) that is capable of inducing genetic mutations.
- *mutualism (mutualistic symbiosis):* a symbiosis in which both partners benefit.
- *natural:* refers to a non-anthropogenic context, i.e., one that is not influenced by humans and is self-organizing and dominated by native species; see also nature.
- *natural capital:* see natural resource
- *natural experiment:* an experiment conducted by observing variations of phenomena in nature, and then developing explanations for these through analysis of potential causal mechanisms. see also, experiment and manipulative experiment.
- *natural gas:* a gaseous, hydrocarbon-rich mixture mined from certain geological formations.
- *natural mortality:* mortality due to natural causes. compare with harvesting mortality.
- *natural population change:* a change in population that is due only to the difference in birth and death rates, and not to immigration or emigration.
- *natural resource:* a source of material or energy that is extracted (harvested) from the environment.
- *natural selection:* a mechanism of evolution, favouring individuals that, for genetically based reasons, are better adapted to coping with environmental opportunities and constraints. these more fit

individuals have an improved probability of leaving descendants, ultimately leading to genetically based changes in populations, or evolution.

- *Nature:* Refers to the entire system of physical and biological existence and organization, uninfluenced by humans; see also natural.
- *Net ecosystem productivity:* The amount of ecosystem-level productivity that remains after respiration is subtracted from gross productivity.
- *Net primary production (NPP):* Primary production that remains as biomass after primary producers have accounted for their respiratory needs. see also, respiration and gross primary production.
- *niche:* the role of a species within its community.
- *nimby:* an acronym for "not in my backyard."
- *nitrification:* the bacterial oxidation of ammonium (NH^{4+}) to nitrate (NO^{3-}).
- *Nitrogen Fixation:* The oxidation of nitrogen gas (N_2) to ammonia (NH_3) or nitric oxide (NO).
- *Noise pollution:* when the level of ambient sound becomes distracting to the normal activities of people. at a higher intensity it can cause hearing impairment.
- *Non-governmental organizations (NGOs):* charities and other not-for-profit organizations. see also, environmental non-governmental organizations.
- *Non-renewable Resource (non-renewable natural resource):* a resource present on earth in finite quantities, so as it is used, its future stocks are diminished. examples are metals and fossil fuels. compare with renewable resources.
- *non-target damage:* damage caused by a pesticide to non-target organisms. see also, broad-spectrum pesticide and non-target organism.
- *non-target organism:* organisms that are not pests, but which may be affected by a pesticide treatment. see also, broad-spectrum pesticide and non-target damage.
- *not in my backyard:* see nimby.

- *nuclear fuel:* unstable isotopes of Uranium (235U) and Plutonium (239PU) that decay through fission, releasing large amounts of energy that can be used to generate electricity.
- *Nuclear winter:* a period of prolonged climate cooling that might be caused by a nuclear war.
- *null hypothesis:* a hypothesis that seeks to disprove a hypothesis.
- *nutrient:* any chemical required for the proper metabolism of organisms.
- *nutrient budget:* a quantitative estimate of the rates of nutrient input and output for an ecosystem, as well as the quantities present and transferred within the system.
- *nutrient capital:* the amount of nutrients present in a site in soil, living vegetation, and dead organic matter.
- *nutrient cycling:* transfers and chemical transformations of nutrients in ecosystems, including recycling through decomposition.
- *ocean:* the largest hydrological compartment, accounting for about 97% of all water on earth.
- *old-growth forest:* a late-successional forest characterized by the presence of old trees, an uneven-aged population structure, and a complex physical structure.
- *oligotrophic:* pertains to aquatic ecosystems that are highly unproductive because of a sparse supply of nutrients. compare with eutrophic and mesotrophic.
- *omnivore:* an animal that feeds on both plant and animal materials.
- *organic agriculture:* systems by which crops are grown using natural methods of maintaining soil fertility, and pest-control methods that do not involve synthetic pesticides.
- *orographic precipitation:* precipitation associated with hilly or mountainous terrain that forces moisture-laden air to rise in altitude and become cooler, causing water vapor to condense into droplets that precipitate as rain or snow.
- *Outer space:* regions beyond the atmosphere of earth.
- *over-harvesting (over-exploitation):* unsustainable harvesting of a potentially renewable resource, leading to a decline of its stocks.

- *oxidizing smog:* an event of air pollution rich in ozone, peroxyacetyl nitrate, and other oxidant gases.
- *paradigm:* a pattern or model; a collection of assumptions, concepts, practices, and values that constitutes a way of viewing reality, especially for an intellectual community that shares them.
- *parameter:* one or more constants that determine the form of a mathematical equation. in the linear equation Y = Ax + B, A and B are parameters, and Y and X are variables. see also, variable.
- *Parasitism:* A biological relationship involving one species obtaining nourishment from a host, usually without causing its death.
- *Peace:* The absence of war.
- *Peace-keeping:* An action that occurs after a hot conflict has stopped through a cease-fire agreement, but the conditions for a lasting peace are not yet in place so various means must be used to keep the antagonists apart. compare with peace-making.
- *peace-making:* the enforced resolution of an active or potential conflict, often by establishing a balanced power relationship among the parties while also imposing a process to achieve a negotiated settlement. compare with peace-keeping.
- *persistence:* the nature of chemicals, especially pesticides, to remain in the environment before eventually being degraded by microorganisms or physical agents such as sunlight and heat.
- *pest:* any organism judged to be significantly interfering with some human purpose.
- *pesticide:* a substance used to poison pests. see also, pest, fungicide, herbicide, and insecticide.
- *pesticide treadmill:* the inherent reliance of modern agriculture and public-health programs on pesticides, often in increasing quantities, to deal with pest problems.
- *Petroleum (crude oil):* a fluid, hydrocarbon-rich mixture mined from certain geological formations.
- *phenotype:* the expressed characteristics of an individual organism, due to genetic and environmental influences on the expression of its specific genetic information. see also, genotype.

- *phenotypic plasticity:* the variable expression of genetic information of an individual, depending on environmental influences during development.
- *photoautotroph:* plants and algae that use sunlight to drive their fixation of energy through photosynthesis. see also, chemoautotroph and photosynthesis.
- *photochemical air pollutants:* ozone, peroxyacetyl nitrate, and other strongly oxidizing gases that form in the atmosphere through complex reactions involving sunlight, hydrocarbons, oxides of nitrogen, and other chemicals.
- *photosynthesis:* autotrophic productivity that utilizes visible electromagnetic energy (such as sunlight) to drive biosynthesis.
- *phytoplankton:* microscopic, photosynthetic bacteria and algae that live suspended in the water of lakes and oceans.
- *plantation:* in forestry, these are tree-farms managed for high productivity of wood fiber.
- *poaching:* the illegal harvesting of wild life (plants or animals).
- *point source:* a location where large quantities of pollutants are emitted into the environment, such as a smokestack or sewer outfall.
- *political ecology:* this integrates the concerns of ecology and political economy to consider the dynamic tensions between natural and anthropogenic change, and also the consideration of damage from both natural and anthropogenic perspectives; the latter includes the broad range of concerns from individual people to all of society.
- *pollution:* the exposure of organisms to chemicals or energy in quantities that exceed their tolerance, causing toxicity or other ecological damages. compare with contamination.
- *population:* in ecology, this refers to individuals of the same species that occur together in time and space.
- *potential energy:* the stored ability to perform work, capable of being transformed into electromagnetic or kinetic energies. potential energy is associated with gravity, chemicals, compressed gases, electrical potential, magnetism, and the nuclear structure of matter.

- *Potentially renewable natural resource:* an alternate phrase for renewable natural resource, highlighting the fact that these can be overexploited, and thereby treated as if they were nonrenewable resources. see also, renewable resource.
- *PPB (Part Per Billion):* A unit of concentration, equivalent to 1 microgram per kilogram (μg/kg), or in aqueous solution, 1 μg per liter (μg/l).
- *PPM (part per million):* a unit of concentration, equivalent to 1 milligram per kilogram (mg/kg), or in aqueous solution, 1 mg per liter (mg/l).
- *Prairie:* grassland ecosystems occurring in temperate regions.
- *precautionary principle:* an approach to environmental management, adopted by many countries at the 1992 earth summit, which essentially states that scientific uncertainty is not a sufficient reason to postpone control measures when there is a threat of harm to human health or the environment.
- *precipitation:* deposition of water from the atmosphere as liquid rain, or as solid snow or hail.
- *precision:* the degree of repeatability of a measurement or observation. compare with accuracy.
- *prevailing wind:* wind that blows in a dominant direction.
- *primary consumer:* a herbivore, or a heterotrophic organism that feeds on plants or algae.
- *primary pollutants:* chemicals that are emitted into the environment. compare with secondary pollutants.
- *primary producer:* an autotrophic organism. autotrophs are the biological foundation of ecological productivity. see also, primary production.
- *primary production:* productivity by autotrophic organisms, such as plants or algae. often measured as biomass accumulated over a unit of time, or sometimes by the amount of carbon fixed.
- *primary sewage treatment:* the initial stage of sewage treatment, usually involving the filtering of larger particles from the sewage wastes, settling of suspended solids, and sometimes chlorination to kill pathogens.

- *principle of limiting factors:* a theory stating that ecological productivity (and some other functions) is controlled by whichever environmental factor is present in least supply relative to the demand.
- *production:* an ecological term related to the total yield of biomass from some area or volume of habitat.
- *production life:* see life index.
- *productivity:* an ecological term for production standardized per unit area and time.
- *prokaryote:* microorganisms without an organized nucleus containing their genetic material. compare with eukaryote.
- *protected area (reserve):* parks, ecological reserves, and other tracts set aside from intense development to conserve their natural ecological values. see also, greater protected area.
- *R-selected:* Refers to organisms that produce relatively large numbers of small offspring. little parental investment is made in each offspring, but having large numbers of progeny helps ensure that some will establish and survive. compare with K-selected.
- *Radiatively Active Gases (RAGs):* Atmospheric gases that efficiently absorb infrared radiation and then dissipate some of the thermal energy gain by reradiation.
- *Rapidly developing countries:* countries with a quickly growing economic infrastructure and a rapidly increasing average per-capita income. see also, high-income and low-income countries and compare with developed countries and less-developed countries.
- *reclamation:* actions undertaken to establish a self-maintaining ecosystem on degraded land, as when a disused industrial site is converted into a permanent cover of vegetation, such as a pasture. compare with restoration and remediation.
- *recycling:* the processing of discarded materials into useful products.
- *relative humidity:* the atmospheric concentration of water, expressed as a percentage of the saturation value for that temperature.
- *remediation:* specific actions undertaken to deal with particular problems of environmental quality, such as the liming of acidic

lakes and rivers to decrease their ecological damage. compare with restoration and reclamation.

- *renewable resource (renewable natural resource):* these can regenerate after harvesting, and potentially can be exploited forever. examples are fresh water, trees, agricultural plants and livestock, and hunted animals. compare with nonrenewable resources.
- *replacement fertility rate:* the fertility rate that results in the numbers of progeny replacing their parents, with no change in size of the equilibrium population.
- *replication:* the biochemical process occurring prior to cellular division, by which information encoded in DNA is copied to produce additional DNA with the same information.
- *Reserve:* (1) Known quantities of resources that can be economically recovered from the environment. (2) an alternative word for a protected area. see protected area.
- *residence time:* (1) the time required for the disappearance of an initial amount; (2) the length of time that a stressor or other environmental influence remains active.
- *residue:* lingering concentrations of pesticides and certain other chemicals in organisms and the environment.
- *resilience:* the ability of a system to recover from disturbance.
- *resistance:* the ability of a population or community to avoid displacement from some stage of ecological development as a result of disturbance or an intensification of environmental stress. changes occur after thresholds of resistance to environmental stressors are exceeded.
- *resource recovery facility:* see waste-to-energy facility.
- *respiration:* physiological processes needed to maintain organisms alive and healthy.
- *response:* in ecotoxicology, this refers to biological or ecological changes caused by exposure to an environmental stressor.
- *restoration:* establishment of a self-maintaining facsimile of a natural ecosystem on degraded land, as when abandoned farmland is converted back to a native prairie or forest. compare with reclamation and remediation.

- *restoration ecology:* activities undertaken by ecologists to repair ecological damage, such as establishing vegetation on degraded habitat, increasing the populations of endangered species, and decreasing the area of threatened ecosystems.
- *reuse:* finding another use for discarded materials, usually with relatively little modification.
- *risk:* see environmental risks.
- *risk assessment:* see environmental risk assessment.
- *rna:* the biochemical ribonucleic acid, which is important in translation of the genetic information of dna into the synthesis of proteins. rna also stores the genetic information of some viruses.
- *ruderal:* short-lived but highly fecund plants characteristic of frequently disturbed environments with abundant resources.
- *run-of-the-river:* a hydroelectric development that directly harnesses the flow of a river to drive turbines, without creating a substantial impoundment for water storage.
- *salinization:* the buildup of soluble salts in the soil surface, an important agricultural problem in drier regions.
- *sanitary landfill:* a facility where municipal solid waste is dumped, compacted by heavy machines, and covered with a layer of clean dirt at the end of the day. some have systems to contain and collect liquid effluent, known as leachate.
- *science:* the systematic and quantitative study of the character and behavior of the physical and biological world.
- *scientific creationist:* a creationist who attempts to explain some of the discrepancies between his or her beliefs (which are based on a literal interpretation of genesis) and scientific understanding of the origin and evolution of life. see also, creationist.
- *scientific method:* this begins with the identification of a question involving the structure or function of the natural world, usually using inductive logic. the question is interpreted in terms of a theory, and hypotheses are formulated and tested by experiments and observations of nature.
- *scrubbing:* see flue-gas desulfurization.
- *Seascape:* A spatial integration of ecological communities over a large marine area.

- *Second law of thermodynamics:* a physical principle stating that transformations of energy can occur spontaneously only under conditions in which there is an increase in the entropy (or randomness) of the universe. see also, first law of thermodynamics and entropy.
- *secondary consumer:* a carnivore that feeds on primary consumers (or herbivores).
- *secondary pollutants:* pollutants that are not emitted, but form in the environment by chemical reactions involving emitted chemicals. compare with primary pollutants.
- *secondary sewage treatment:* treatment applied to the effluent of primary sewage treatment, usually involving the use of a biological technology to aerobically decompose organic wastes in an engineered environment. the resulting sludge can be used as a soil conditioner, incinerated, or dumped into a landfill. see also, primary sewage treatment.
- *sedimentary rock:* rock formed from precipitated minerals such as calcite, or from lithified particles eroded from other rocks such as sandstone, shale, and conglomerates.
- *sedimentation:* a process by which mass eroded from elsewhere settles to the bottom of rivers, lakes, or an ocean.
- *seismic sea wave:* see Tsunami.
- *Selection harvesting:* harvesting of only some trees from a stand, leaving others behind and the forest substantially intact.
- *sewage treatment:* the use of physical filters, chemical treatment, and/or biological treatment to reduce pathogens, organic matter, and nutrients in waste waters containing sewage.
- *shifting cultivation:* an agricultural system in which trees are felled, the woody debris burned, and the land used to grow mixed crops for several years.
- *significant figures:* the number of digits used when reporting data from analyses or calculations.
- *silvicultural management:* the application of practices that increase tree productivity in a managed forest, such as planting seedlings, thinning trees, or applying herbicides to reduce the abundance of weeds.

- *silviculture:* the branch of forestry concerned with the care and tending of trees.
- *site capability (site quality):* the potential of land to sustain the productivity of agricultural crops.
- *slash-and-burn:* an agricultural system that results in a permanent conversion of a forest into crop production, involving cutting and burning the forest followed by continuous use of the land for crops.
- *slope:* the angle of inclination of land, measured in degrees (0° implies a horizontal surface, while 90° is vertical).
- *Sloss:* An acronym, for single large or several small, in reference to choices in the design of protected areas.
- *Sludge:* A solid or semi-solid precipitate that settles from polluted water during treatment; sludge is produced during the treatment of sewage and also in pulp mills and some other industrial facilities. it may be disposed of in a landfill, but if organic, can be used as a beneficial soil amendment.
- *smog:* an event of ground-level air pollution.
- *snag:* a standing dead tree.
- *social justice:* a worldview that calls for equality of consideration for all members of a society, regardless of color, race, socio-economic class, gender, age, or sexual preference. see also, ecological justice.
- *socio-cultural evolution:* see cultural evolution.
- *socio-economic impact assessment:* a process used to identify and evaluate the potential consequences of proposed actions or policies for sociological, economic, and related values. see also, environmental impact assessment.
- *soil:* a complex mixture of fragmented rock, organic matter, moisture, gases, and living organisms that covers almost all of earth's terrestrial landscapes.
- *soil profile:* the vertical stratification of soil on the basis of color, texture, and chemical qualities.
- *solar energy:* electromagnetic energy radiated by the sun.
- *solar system:* the sun, its nine orbiting planets, miscellaneous comets, meteors, and other local materials.

- *solid wastes:* extremely variable municipal wastes that include discarded food, garden discards, newspapers, bottles, cans, construction debris, old cars, and disused furniture.
- *spaceship earth:* an image of earth as viewed from space, which illustrates the fact that, except for sunlight, resources needed by humans are present only on that planet.
- *spaceship world view:* this focuses on sustaining only those resources needed by humans and their economy, and it assumes that humans can exert a great degree of control over natural processes and can pilot "spaceship earth." see also frontier world view and sustainability world view.
- *special concern:* refers to a species that is not currently threatened but is at risk of becoming so for various reasons.
- *species:* an aggregation of individuals and populations that can potentially interbreed and produce fertile offspring, and is reproductively isolated from other such groups.
- *speciesism:* discrimination (by humans) against other species purely on the basis that they are not human, especially as manifested by cruelty to or exploitation of animals, or merely by a lack of consideration of their interests.
- *species richness:* the number of species in some area or place.
- *State-of-the-environment Reporting:* A governmental, corporate, or ngo function that involves public reporting on environmental conditions.
- *strategic weapon:* large explosive-yield weapons that are designed to be delivered by a missile or airplane over a distance of thousands of kilometers. compare with tactical weapon.
- *stratosphere:* the upper atmosphere, extending above the. from 8–17 km to as high as about 50 km. see also, troposphere.
- *stress-tolerator:* long-lived plants adapted to habitats that are marginal in terms of climate, moisture, or nutrient supply, but are infrequently disturbed and therefore stable, such as tundra and desert.
- *stressor:* an environmental factor that constrains the development and productivity of organisms or ecosystems.
- *succession:* a process of community- level recovery following disturbance.

- *surface flow:* water that moves over the surface of the ground.
- *surface water:* water that occurs in glaciers, lakes, ponds, rivers, streams, and other surface bodies of water.
- *sustainability world view:* this acknowledges that humans must have access to vital resources, but it asserts that the exploitation of resources should be governed by appropriate ecological, aesthetic, and moral values, and should not deplete the necessary resources. see also, frontier world view and spaceship world view.
- *sustainable development:* refers to progress toward an economic system that uses natural resources in ways that do not deplete their stocks or compromise their availability to future generations.
- *sustainable economic system (sustainable economy):* an economic system that can be maintained over time without any net consumption of natural resources.
- *swamp:* a forested wetland, flooded seasonally or permanently.
- *symbiosis:* an intimate relationship between different species. see also, mutualism.
- *synecology:* the study of relationships among species within communities. compare with autecology.
- *system:* a group or combination of regularly interacting and interdependent elements, which form a collective entity, but one that is more than the sum of its constituents. see also, ecosystem.
- *tactical weapon:* smaller, numerous weapons that are intended for use in a local battlefield, and are delivered by smaller missiles, artillery, aircraft, or torpedoes. compare with strategic weapon.
- *taiga:* see boreal forest.
- *Tectonic force:* force associated with crustal movements and related geological processes that cause structural deformations of rocks and minerals.
- *temperate deciduous forest:* a forest occurring in relatively moist, temperate climates with short and moderately cold winters and warm summers, and usually composed of a mixture of angiosperm tree species.
- *temperate grassland:* grass-dominated ecosystems occurring in temperate regions with an annual precipitation of 25–60 cm per year; sufficient to prevent desert from developing but insufficient to support forest.

- *temperate rainforest:* a forest developing in a temperate climate in which winters are mild and precipitation is abundant year-round. because wildfire is rare, old-growth forests may be common.
- *temperature inversion:* see atmospheric inversion.
- *teratogen:* a chemical or physical agent that induces a developmental abnormality (i.e., a birth defect) in an embryo or foetus.
- *tertiary sewage treatment:* treatment applied to the effluent of secondary sewage treatment, usually involving a system to remove phosphorus and/or nitrogen from waste waters. see also, primary sewage treatment and secondary sewage treatment.
- *theory:* a general term that refers to a set of scientific laws, rules, and explanations supported by a large body of experimental and observational evidence, all leading to robust, internally consistent conclusions.
- *thermal pollution:* an increase in environmental temperature sufficient to result in ecological change.
- *thermosphere:* the layer of the atmosphere extending beyond the mesosphere to 450 km or more above the surface of the earth. see also, mesosphere.
- *threatened:* in canada, this refers to any indigenous taxa likely to become endangered (in canada) if factors affecting their vulnerability are not reversed.
- *tidal energy:* energy that develops in oceanic surface waters because of the gravitational attraction between earth and the moon, and can potentially be used to generate electricity.
- *tilth:* the physical structure of soil, closely associated with the concentration of humified organic matter. tilth is important in water- and nutrient-holding capacity of soil, and is generally beneficial to plant growth.
- *tolerance:* in ecotoxicology, this refers to a genetically based ability of organisms or species to not suffer toxicity when exposed to chemicals or other stressors.
- *Total concentration:* the concentrations of metals in soil, sediment, rocks, or water, as determined after dissolving samples in a strongly acidic solution. compare with available concentration.

- *Total energy production:* the use of commercial energy plus traditional fuels in an economy. see also, traditional fuels.
- *total-war economy:* an economy that is wholly dedicated to supporting a war effort.
- *toxic waste:* waste that is poisonous to humans, animals, or plants. see also, hazardous waste.
- *toxicology:* the science of the study of poisons, including their chemical nature and their effects on the physiology of organisms. compare with environmental toxicology and ecotoxicology.
- *traditional fuels:* the non-commercial use of wood, charcoal, animal dung, and other biomass fuels for subsistence purposes, primarily for cooking food and heating homes. see also, total energy production and commercial energy production.
- *transcription:* a biochemical process by which the information of double-stranded dna is encoded on complementary single strands of RNA, which are used to synthesize specific proteins.
- *Translation:* A biochemical process occurring on organelles known as ribosomes, in which information encoded in messenger RNA is used to synthesize particular proteins.
- *Transpiration:* The evaporation of water from plants. compare with evapotranspiration.
- *trophic structure:* the organization of productivity in an ecosystem, including the roles of autotrophs, herbivores, carnivores, and detritivores.
- *troposphere:* the lower atmosphere, extending to 8–17 km.
- *tsunami:* a fast-moving, sea-wave caused by undersea earthquakes, which if large can cause enormous destruction of low-lying coastal places.
- *tundra:* a treeless biome occurring in environments with long, cold winters and short, cool growing seasons.
- *Unexploded ordinance (UXOS):* Explosives that remain in place after a conflict has ended.
- *Urban agglomeration:* see megacity.
- *urban forest:* urban areas having a substantial density and biomass of trees, although often most are non-native species.

- *urban planning:* an active process of designing better ways of organizing the structure and function of cities, including an orderly siting of land uses and activities.
- *urbanization:* the development of cities and towns on formerly agricultural or natural lands.
- *utilitarian value:* the usefulness of a thing or function to humans.
- *value added:* the increased value of something as a result of manufacturing or some other improvement.
- *valued ecosystem components (VECS):* In environmental impact assessment, these are components of ecosystems perceived to be important to society as economically important resources, as rare or endangered species or communities, or for their cultural or aesthetic significance.
- *Variable:* A changeable factor believed to influence a natural phenomenon of interest or that can be manipulated during an experiment.
- *vascular plant:* relatively complex plants with specialized, tube-like vascular tissues in their stems for conducting water and nutrients.
- *VECS:* See valued ecosystem components.
- *Vector:* Species of insects and ticks that transmit pathogens from alternate hosts to people or animals.
- *vertebrate:* animals with an internal skeleton, and in particular a backbone.
- *virgin field:* in epidemiology, this is a population that is hypersensitive to one or more infectious diseases to which it has not been previously exposed.
- *volatile organic compounds:* organic compounds that evaporate to the atmosphere at typical environmental temperatures, so they are present in gaseous or vapor forms.
- *volcano:* an opening in earth's crust from which magmatic materials, such as lava, rock fragments, and gases, are ejected into the atmosphere or oceanic waters.
- *war:* a period of organized deadly conflict between human societies, countries, or another defined group.

- *waste:* any discarded materials. see also, hazardous waste and toxic waste.
- *waste management:* the handling of discarded materials using various methods. see also, dumping, incineration, recycling, composting, reuse, and waste reduction.
- *waste prevention:* see waste reduction.
- *waste reduction:* practices intended to reduce the amount of waste that must be disposed of. also known as waste prevention.
- *waste-to-energy facility:* an incinerator that burns organic waste and uses the heat generated to produce commercial energy.
- *water cycle:* see hydrologic cycle.
- *watershed:* an area of land from which surface water and groundwater flow into a stream, river, or lake.
- *wave energy:* the kinetic energy of oceanic waves, which can be harnessed using specially designed buoys to generate electricity.
- *weather:* the short-term, day-to-day or instantaneous meteorological conditions at a place or region. compare with climate.
- *weathering:* physical and chemical processes by which rocks and minerals are broken down by such environmental agents as rain, wind, temperature changes, and biological influences.
- *weed:* an unwanted plant that interferes with some human purpose.
- *wet deposition:* atmospheric inputs of chemicals with rain and snow. compare with dry deposition.
- *wetland:* an ecosystem that develops in wet places and is intermediate between aquatic and terrestrial ecosystems. see also, bog, fen, marsh, and swamp.
- *whole-lake experiment:* the experimental manipulation of one or more environmental factors in an entire lake.
- *wind:* an air mass moving in earth's atmosphere.
- *wind energy:* the kinetic energy of moving air masses, which can be tapped and utilized in various ways, including the generation of electricity.
- *work:* in physics, work is defined as the result of a force being applied over a distance.

- *working hypothesis:* a hypothesis being tested in a scientific experiment or another kind of research. see also, hypothesis and null hypothesis.
- *Zero population growth (ZPG):* When the birth rate plus immigration equal the death rate plus emigration.
- *zooplankton:* tiny animals that occur in the water column of lakes and oceans.

Bibliography

1. Abrams, Peter. (1992). Resource. In Evelyn, F. Keller and Elisabeth, A. Lloyd, editors, *Keywords in Evolutionary Biology*. Cambridge: Harvard University Press.
2. Allaby, M. (1998). *Oxford Dictionary of Ecology*. New York, NY: Oxford University Press.
3. Anton, John, P. (1999). Aristotle and Theophrastus on Ecology. In, K. Boudoris and, K. Kalimtzis, editors, *Philosophy and Ecology.* Volume 1. Athens, Greece: Ionia Publications.
4. Art, H. W. (1993). *The Dictionary of Ecology and Environmental Science*. New York, NY: Henry Holt and Company.
5. Beisner, Beatrix and Kim Cuddington, editors. (2005). *Ecological Paradigms Lost: Routes of Theory Change.* Academic Press.
6. Blackstone, William, T. (1973). Ethics and Ecology. *Southern Journal of Philosophy 11,* 55–71.
7. Booth, B. D., S. D. Murphy, & C. J. Swanton. (2003). *Weed Ecology in Natural and Agricultural Systems.* Cambridge, MA: CABI Publishing.
8. Boucher, Douglas, H. (1992). Mutualism and Cooperation. In Evelyn, F. Keller and Elisabeth, A. Lloyd, editors, *Keywords in Evolutionary Biology*. Cambridge: Harvard University Press.
9. Buege, Douglas, J. (1996). An Ecologically-Informed Ontology for Environmental Ethics. *Biology & Philosophy 12(1),* 1–20.
10. Carlton, J.T. (2001). Introduced Species in U.S. Coastal Waters: Pew

Oceans Commissions Report. Pew Oceans Commissions: Washington, DC.

11. Castle, David, G. A. (1997). Three Applications of the Semantic Conception of Theories to Problems in Ecology. Doctoral dissertation, University of Guelph.
12. Castle, David, G. A. (2001b). A Semantic View of Ecological Theories. Dialectica: *International Journal of Philosophy of Knowledge 55(1),* 51–65.
13. Castle, David. (2000). Limitations on an Inclusive Definition of Ecosystem-Human Health. *Ethics and the Environment 5,* 153–161.
14. Castle, David. (2001a). A Gradualist Theory of Discovery in Ecology. *Biology & Philosophy 16(4),* 547–571.
15. Clements, Frederic, E., & Victor, E. Shelford. (1939). *Bio-ecology.* John Wiley & Sons, New York. 425 pp.
16. Cohen, A. H., & J. T. Carlton. (1998). Accelerating invasion rate in a highly invaded estuary. *Science 279,* 555–558.
17. Colautti, R. I., & H. J. MacIsaac. (2004). A neutral terminology to define 'invasive' species. *Diversity and Distributions 10,* 134–141.
18. Colwell, Robert, K. (1992). Niche: A Bifurcation in the Conceptual Lineage of the Term. In Evelyn, F. Keller and Elisabeth, A. Lloyd, editors, *Keywords in Evolutionary Biology.* Cambridge: Harvard University Press.
19. Colyvan, Mark and, L. Ginzburg. (2003a). The Gallilean Turn in Population Ecology. *Biology & Philosophy 18(3),* 401–414.
20. Colyvan, Mark and, L. R. Ginzburg. (2003b). Laws of nature and laws of ecology. *Oikos* 10*1(3),* 649.
21. Colyvan, Mark. (2005). Probability and Ecological Complexity. *Biology & Philosophy 20(4),* 869–879.
22. Colyvan, Mark. 2023. Population Ecology. In Sahotra Sarkar and Anya Plutynski, editors, A Companion to the *Philosophy of Biology.* Blackwell.
23. Cooper, Gregory, J. (1989). Ecology and evolution: a philosophical study. Ph.D. thesis, University of Minnesota.
24. Cooper, Gregory, J. (2004). The Science of the Struggle for Existence: On the Foundations of Ecology. Cambridge: Cambridge University Press.

25. Cooper, Gregory. (1990). The Explanatory Tools of Theoretical Population Ecology. In *PSA* 1990, S165–S178.
26. Cooper, Gregory. (1993). The Competition Controversy in Community Ecology. *Biology & Philosophy 8,* 359–384.
27. Cooper, Gregory. (1996). Theoretical Modeling and Biological Laws. *Philosophy of Science* (Proceedings, Vol.1): S29–S35.
28. Cooper, Gregory. (1998). Generalizations in Ecology: A Philosophical Taxonomy. *Biology & Philosophy 13,* 555–586.
29. Cooper, Gregory. (2001). Must There Be a Balance of Nature? *Biology & Philosophy 16(4),* 481–506.
30. Cuddington, Kim. (2001). The `Balance of Nature' Metaphor and Equilibrium in Population Ecology. *Biology & Philosophy 16(4),* 463–479.
31. de Laplante, Kevin. (1998). Toward a General Philosophy of Ecology. Ph.D. thesis, University of Western Ontario.
32. de Laplante, Kevin. (2004). Toward a more expansive conception of ecological science. *Biology & Philosophy 19,* 263–281.
33. Delord, Julien. (2006). Paths Toward a Proper Philosophy of Ecology. *Biological Theory 1(4),* 423–427.
34. Dupré, John. (1993). Reductionism in Biology I: Ecology. In *The Disorder of Things: Metaphysical Foundations of the Disunity of Science.* Cambridge: Harvard University Press.
35. Egerton, Frank, N. (1993). The history and present entanglements of some general ecological perspectives. In Steward, T.A. Pickett and Mark, J. McDonnell, editors, *Humans as Components of Ecosystems,* 9–23. New York: Springer-Verlag.
36. Eliot, Christopher. (2004). Exceptions Make the Rules: The Role of Disturbing Conditions in Ecological Theorizing. Ph.D. thesis, University of Minnesota.
37. Eliot, Christopher. (2007). Method and Metaphysics in Clements's and Gleason's Ecological Explanations. *Studies in History and Philosophy of Biological and Biomedical Sciences 38(1),* 85–109.
38. Elliot, Robert. (1994). Ecology and the Ethics of Environmental Restoration. *Philosophy* 36(Supp): 31–43.
39. Faith, Daniel, P. (2003). Biodiversity. In *Stanford Encyclopedia of Philosophy.* http://plato.stanford.edu/entries/biodiversity/, The Metaphysics Research Lab, Stanford University.

40. Flew, A. G. N. (1962). The Structure of Malthus' Population Theory. In Bernard Baumrin, editor, *Philosophy of Science: The Delaware Seminar*, Volume 1, 1961–1962. New York: Interscience Publishers.
41. Garfinkel, Alan. (1981). Reductionism. In *Forms of Explanation.* New Haven: Yale University Press.
42. Garson, J., & James Justus. (2006). Conservation Biology. In Sahotra Sarkar and Jessica Pfeifer, editors, *The Philosophy of Science: An Encyclopedia*. Routledge.
43. Ginzburg, L. and, M. Colyvan. (2004). *Ecological Orbits: How Planets Move and Populations Grow.* Oxford: Oxford University Press.
44. Grene, Marjorie. (1980). A Note on Simberloff's `Succession of Paradigms in Ecology.' In Esa Saarinen, editor, *Conceptual Issues in Ecology*. Boston: Dordrecht Reidel.
45. Griesemer, James, R. (1988). Causal Explanation in Laboratory Ecology: The Case of Competitive Indeterminacy. In *PSA* 1988, S337–S344.
46. Griesemer, James, R. (1992). Niche: Historical Perspectives. In Evelyn, F. Keller and Elisabeth, A. Lloyd, editors, *Keywords in Evolutionary Biology*. Cambridge: Harvard University Press.
47. Hagen, Joel, B. (1989). Research Perspectives and the Anomalous Status of Modern Ecology. *Biology & Philosophy 4,* 433–455.
48. Haila, Yrjö and Olli Järvinen. (1982). The Role of Theoretical Concepts in Understanding the Ecological Theatre: A Case Study on Island Biogeography. In Esa Saarinen, editor, *Conceptual Issues in Ecology.* Dordrecht: Riedel.
49. Haila, Yrjö and Richard Levins. (1992). *Humanity and Nature: Ecology, Science, and Society.* London: Pluto Press.
50. Haila, Yrjö. (1986). On the Semiotic Dimension of Ecological Theory: The Case of Island Biogeography. *Biology & Philosophy 1,* 377–387.
51. Haila, Yrjö. (1988). The Multiple Faces of Ecological Theory and Data. *Oikos 53,* 408–411.
52. Haila, Yrjö. (1989). Ecology Finding Evolution Finding Ecology. *Biology & Philosophy 4,* 235–244.
53. Haila, Yrjö. (1995). Natural Dynamics as a Model for Management: Is the Analogue Practicable? In, A. L. Sippola, P. Alaraudanjoki, B. Forbes, & V. Hallikainen, editors, *Northern Wilderness Areas: Ecology, Sustainability, Values. Rovaniemi*: Arctic Center Publications.

54. Haila, Yrjö. (1997). A `Natural' Benchmark for Ecosystem Function. *Conservation Biology 11,* 300–301.

55. Haila, Yrjö. (2000). Beyond the Nature-Culture Dualism. *Biology & Philosophy 15(2),* 155–175.

56. Heutte, T., & E. Bella. (2003). Invasive plants and exotic weeds of Southeast Alaska. Anchorage, AK: USDA Forest Service.

57. Hourdequin, Marion. (2005). Theories as tools: a pluralistic approach to ecological modeling. *Studies in History and Philosophy of Biological and Biomedical Sciences 36(3),* 594–601.

58. Ives, Anthony, R. (2005). Community Diversity and Stability: Changing Perspectives and Changing Definitions. In Beisner and Cuddington (2005), 159–182.

59. Jeschke, J. M., & D. L. Strayer. (2005). Invasion success of vertebrates in Europe and North America. *Proceedings of the National Academy of Sciences 102*(20):7198–202.

60. Justus, James and Sahotra Sarkar. (2002). The Principle of Complementarity in the Design of Reserve Networks to Conserve Biodiversity: A Preliminary History. *Journal of Biosciences 27,* 421–435.

61. Justus, James. (2002). Taking Stock of Conservation Biology: Old Problems and New Directions. *Bioscience 52,* 854–856.

62. Justus, James. (2004). Review of Searching for Sustainability: Interdisciplinary Essays in the Philosophy of Conservation Biology. *Philosophy of Science 71,* 232–235.

63. Justus, James. (2023a, in press). Complexity, Diversity, Stability. In Sahotra Sarkar and Anya Plutynski, editors, A Companion to the *Philosophy of Biology.* Malden, Mass.: Blackwell.

64. Justus, James. (2023b, in press). Ecological and Lyapunov Stability. *Philosophy of Science.*

65. Justus, James. (2023c, in press). Loop Analysis and Qualitative Modeling: Limitations and Merits. *Biology & Philosophy.*

66. Justus, James. (2023d, in press). Qualitative Scientific Modeling and Loop Analysis. *Philosophy of Science.*

67. Keller, David, R., & Frank, B. Golley. (2000). *Science of Synthesis: An Introduction to the Philosophy of Ecology.* University of Georgia Press.

68. Keller, Evelyn, F. (1992). Competition: Current Usages. In Evelyn, F. Keller and Elisabeth, A. Lloyd, editors, *Keywords in Evolutionary Biology*. Cambridge: Harvard University Press.
69. Kirkman, Robert. (1997). Why Ecology Cannot Be All Things to All People. *Environmental Ethics 19(4),* 375–390.
70. Leibold, M. A., & Gregory, M. Mikkelson. (2002). Coherence, species turnover, and boundary clumping: Elements of meta-community structure. *Oikos 97,* 237–250.
71. Levins, Richard and Richard Lewontin. (1980). Dialectics and Reductionism in Ecology. In Esa Saarinen, editor, *Conceptual Issues in Ecology*. Boston: Dordrecht Reidel.
72. Lloyd, Elisabeth. (1987). Confirmation of Ecological and Evolutionary Models. *Biology & Philosophy 2,* 277–293.
73. Lodge, David and Kristin S Shrader-Frechette. (2002). Nonindigenous Species: Ecological Explanation, Environmental Ethics, & Public Policy. *Conservation Biology 17(1),* 31–37.
74. Loehle, Craig. (1988). Philosophical Tools: Potential Contributions to Ecology. *Oikos 51(1),* 97–104.
75. Loehle, Craig. (1990). Philosophical Tools: Reply to Shrader-Frechette and McCoy. *Oikos 58,* 115–119.
76. Looijen, Richard Christiaan. (1998). The Reduction Problem in Ecology. In *Holism and Reductionism in Biology and Ecology: The Mutual Dependence of Higher and Lower Level Research Programmes*. University of Groningen.
77. Macdonald, Helen. (2002). 'What makes you a scientist is the way you look at things': ornithology and the observer 1930–1955. *Studies in History and Philosophy of Biological and Biomedical Sciences 33,* 53–77.
78. McIntosh, Robert, P. (1992a). Competition: Historical Perspectives. In Evelyn, F. Keller and Elisabeth, A. Lloyd, editors, *Keywords in Evolutionary Biology*. Cambridge: Harvard University Press.
79. McIntosh, Robert, P. (1992b). Whither Ecology? A review of A Critique for Ecology, By Robert Henry Peters. *Quarterly Review of Biology 67(4),* 495–498.
80. Mikkelson, Gregory, M. (1996a). The Large, the Small, the Real, and

the Natural. *Biology & Philosophy 11,* 127–132.

81. Mikkelson, Gregory, M. (1996b). Stretched Lines, Averted Leaps, & Excluded Competition: A Theory of Scientific Counterfactuals. In *PSA* 1996, vol. 63, S194–S201.
82. Mikkelson, Gregory, M. (1997a). Methods and Metaphors in Community Ecology: The Problem of Defining Stability. *Perspectives on Science 5,* 481–498.
83. Mikkelson, Gregory, M. (1997b). Other Things Being Equal: Counterfactuals, Natural Laws, & Scientific Models; with Case Studies from Ecology. Ph.D. thesis, University of Chicago.
84. Mikkelson, Gregory, M. (1998). Counterfactuals, Experimental Reasoning, & Ecology. Communiqué, Canadian Society for the History and *Philosophy of Science* 41(14).
85. Mikkelson, Gregory, M. (2001a). Complexity and Verisimilitude: Realism for Ecology. *Biology & Philosophy 16,* 533–546.
86. Mikkelson, Gregory, M. (2001b). Untangling Ecology? *Biology & Philosophy 16,* 273–279.
87. Mikkelson, Gregory, M. (2003). Ecological Kinds and Ecological Laws. *Philosophy of Science 70,* 1390–1400.
88. Mikkelson, Gregory, M. (2004). The Science of the Struggle for Existence. *Notre Dame Philosophical Reviews.*
89. Mikkelson, Gregory, M. (2023a, in press). Ecological Economics. In Michael Ruse, editor, *Philosophy of Biology.* Prometheus Books.
90. Mikkelson, Gregory, M. (2023b, in press). Ecology. In David, L. Hull and Michael Ruse, editors, *Cambridge Companion to the Philosophy of Biology.* Cambridge Univ. Press.
91. Mikkelson, Gregory, M. (2023c, in press). Ecology as a Moral Science. In Jr. Skipper, Robert, A., Colin Allen, Rachel, A. Ankeny, Carl, F. Craver, Lindley Darden, Greg Mikkelson, & Robert Richardson, editors, *Philosophy and the Life Sciences: A Reader*. Cambridge, Mass.: MIT Press.
92. Mikkelson, Gregory, M. (2023d, in press). Part-Whole Relationships and the Unity of Ecology. In Jr. Skipper, Robert, A., Colin Allen, Rachel, A. Ankeny, Carl, F. Craver, Lindley Darden, Greg Mikkelson, & Robert Richardson, editors, *Philosophy and the Life Sciences: A Reader*. Cambridge, Mass.: MIT Press.

93. Mikkelson, Gregory, M. (2023e, in press). Realism vs. Instrumentalism in a New Statistical Framework. *Philosophy of Science.*
94. Mikkelson, Gregory, M. 20. Niche-based vs. Neutral Models of Ecological Communities. *Biology & Philosophy* 557–566.
95. Noss, R. F., & A. Y. Cooperrider. (1994). Saving Nature's Legacy: Protecting and Restoring Biodiversity. Washington, DC: Island Press.
96. Odenbaugh, Jay and Kevin de Laplante. (2023). What Isn't Wrong with Ecosystem Ecology. In Robert, A. Skipper, Jr., Colin Allen, Rachel, A. Ankeny, Carl, F. Craver, Lindley Darden, Greg Mikkelson, & Robert Richardson, editors, *Philosophy and the Life Sciences: A Reader*. MIT Press.
97. Odenbaugh, Jay. (2001a). Ecological Stability, Model Building, & Environmental Policy: A Reply to Some of the Pessimism. *Philosophy of Science 68*(Proceedings): S493–S505.
98. Odenbaugh, Jay. (2001b). *Searching for Patterns, Hunting for Causes: A Philosophical Examination of Mathematical Modeling in Theoretical Ecology.* PhD thesis, University of Calgary.
99. Odenbaugh, Jay. (2003). Advocacy, Value, & Conservation Biology. *Environmental Values 12(1),* 55–69.
100. Odenbaugh, Jay. (2005a). Idealized, Inaccurate but Successful: A Pragmatic Approach to Evaluating Models in Theoretical Ecology. *Biology & Philosophy 20*(2–3): 231–255.
101. Odenbaugh, Jay. (2005b). The Structure of Population Ecology: Philosophical Reflections on Unstructured and Structured Models. In Beatrix Beisner and Kim Cuddington, editors, *Ecological Paradigms Lost: Routes of Theory Change*. Academic Press.
102. Odenbaugh, Jay. (2006a). Ecology. In Sahotra Sarkar and Jessica Pfeifer, editors, *The Philosophy of Science: An Encyclopedia*. Routledge.
103. Odenbaugh, Jay. (2006b). Struggling with the Sciences of Ecology. *Biology & Philosophy 21,* 395–409.
104. Odenbaugh, Jay. (2023a, in press). A Message in the Bottle? The Constraints of Experimentation on Scientific Reasoning. *Philosophy of Science.*
105. Odenbaugh, Jay. (2023b, in press). Models. In *Blackwell Companion to the Philosophy of Biology.* Blackwell.
106. Odenbaugh, Jay. (2023c, in press). Seeing the Forest and the Trees. *Philosophy of Science.*

107. Odenbaugh, Jay. (2023d, in press). The Strategy of 'The Strategy of Model Building in Population Biology.' *Biology & Philosophy.*

108. Odenbaugh, Jay. 2023. Complex Systems, Trade-Offs and Mathematical Modeling: A Response to Sober and Orzack. *Philosophy of Science.*

109. Odum, E. P. (1959). *Fundamentals of Ecology*. W. B. Saunders Co., Philadelphia and London. 546 p.

110. Orzack, Steven, H., & Elliott Sober. (1993). A Critical Assessment of Levins's The Strategy of Model Building in Population Biology (1966). *Quarterly Review of Biology 68(4),* 533–546.

111. Pickett, Steward, T. A., Jurek Kolasa, & Clive Jones. (1994). *Ecological Understanding: The Nature of Theory and the Theory of Nature*. San Diego: Academic Press.

112. Pietarinen, Juhani. (1999). Plato and Biodiversity. In K. Boudoris and, K. Kalimtzis, editors, *Philosophy and Ecology*, vol. 1. Athens: Ionia Publications.

113. Rapport, David, J. (1995). Ecosystem Health: More than a Metaphor? *Environmental Values 4(4),* 287–309.

114. Richardson, Robert, C. (1994). Optimization in Evolutionary Ecology. In *PSA* 1994, vol. 1, S13–S21.

115. Royer, F., & R. Dickinson. (1999). *Weeds of the Northern*, U.S., & Canada. Edmonton, AB: Lone Pine Press.

116. Sagoff, Mark. (1984). On Explanation in Ecology. University of Maryland Center for Philosophy and Public Policy working paper.

117. Sagoff, Mark. (1985). Fact and Value in Ecological Science. *Environmental Ethics 7,* 99–116.

118. Sagoff, Mark. (2003). The Plaza and the Pendulum: Two Concepts of Ecological Science. *Biology & Philosophy 18(4),* 529–552.

119. Sarkar, Sahotra, James Justus, T. Fuller, C. Kelley, J. Garson, & M. Mayfield. (2005). Effectiveness of Estimator Surrogates for the Selection of Conservation Area Networks. *Conservation Biology 19,* 815–825.

120. Sarkar, Sahotra. (1984). review of Hierarchy: Perspectives for Ecological Complexity. *Philosophy of Science 51,* 359–361.

121. Sarkar, Sahotra. (1996). Ecological Theory and Anuran Declines. *Bioscience 46,* 199–207.

122. Sarkar, Sahotra. (2004). Conservation Biology. In Edward, N. Zalta,

editor, *Stanford Encyclopedia of Philosophy.* http://plato.stanford.edu/entries/conservation-biology/, The Metaphysics Research Lab, Stanford University.

123. Sarkar, Sahotra. (2005). Ecology. In Edward, N. Zalta, editor, *Stanford Encyclopedia of Philosophy.* http://plato.stanford.edu/entries/ecology/, The Metaphysics Research Lab, Stanford University.
124. Shrader Frechette, Kristin and Earl, D. McCoy. (1994). How the Tail Wags the Dog: How Value Judgments Determine Ecological Science. *Environmental Values 3(2),* 107–120.
125. Shrader-Frechette, Kristin, S. (1986). Organismic Biology and Ecosystems Ecology: Description or Explanation? In, N. Rescher, editor, *Current Issues in Teleology.* Pittsburgh: University of Pittsburgh Press.
126. Shrader-Frechette, Kristin, S. (1989a). Ecological Theories and Ethical Imperatives. In, W. R. Shea and, B. Sitter, editors, *Scientists and Their Responsibility*. Canton: Watson.
127. Shrader-Frechette, Kristin, S. (1989b). Scientific Method and the Objectivity of Epistemic Value Judgments. In, J. Fenstad, R. Hilpinen, & I. Frolov, editors, *Logic, Methodology, and the Philosophy of Science.* New York: Elsevier Science Publishers.
128. Shrader-Frechette, Kristin, S. (1990a). Interspecific Competition, Evolutionary Epistemology, & Ecology. In, N. Rescher, editor, *Evolution, Cognition, and Realism.* Lanham: University Press of America.
129. Shrader-Frechette, Kristin, S. (1990b). Island Biogeography, Species-Area Curves, and Statistical Errors: Applied Biology and Scientific Rationality. In *PSA* 1990, S447–S456.
130. Shrader-Frechette, Kristin, S. (1990c). Models, Scientific Method, and Environmental Ethics. In, D. Scherer, editor, *Upstream/Downstream: Issues in Environmental Ethics*. Philadelphia: Temple University Press.
131. Shrader-Frechette, Kristin, S. (1994a). Ecological Explanation and the Population-Growth Thesis. In *PSA* 1994, vol. 1, S34–S45.
132. Shrader-Frechette, Kristin, S. (1994b). Ecosystem Health: A New Paradigm for Ecological Assessment? *Trends in Ecology and Evolution 9*(12), 456–457.
133. Shrader-Frechette, Kristin, S. (1995a). Hard Ecology, Soft Ecology, and Ecosystem Integrity. In Laura Westra and John Lemons, editors,

Perspectives on Ecological Integrity. Boston: Kluwer Academic Publishers.

134. Shrader-Frechette, Kristin, S. (1995b). Practical Ecology and Foundations for Environmental Ethics. *Journal of Philosophy 92*(12), 621–635.
135. Shrader-Frechette, Kristin, S. (1997). Ecological Theory is Problematic. In Byron Lamont, editor, *Environmental Biology.* Perth: Curtin University of Technology.
136. Shrader-Frechette, Kristin, S. (2000). Ecology. In Dale Jamieson, editor, *A Companion to Environmental Philosophy.* Cambridge: Blackwell.
137. Shrader-Frechette, Kristin, S., & Earl, D. McCoy. (1990). Theory reduction and explanation in ecology. *Oikos 58(1),* 109–114.
138. Shrader-Frechette, Kristin, S., & Earl, D. McCoy. (1992a). Community Ecology, Scale, and the Instability of the Stability Concept. In *PSA* 1992, vol. 1, S184–S199.
139. Shrader-Frechette, Kristin, S., & Earl, D. McCoy. (1992b). Statistics, Costs and Rationality in Ecological Inference. *Trends in Ecology and Evolution 7(3),* 96–99.
140. Shrader-Frechette, Kristin, S., & Earl, D. McCoy. (1993). *Method in Ecology*. Cambridge: Cambridge University Press.
141. Shrader-Frechette, Kristin, S., & Earl, D. McCoy. (1994a). Applied Ecology and the Logic of Case Studies. *Philosophy of Science 61,* 228–249.
142. Shrader-Frechette, Kristin, S., & Earl, D. McCoy. (1994b). Biodiversity, Biological Uncertainty, & Setting Conservation Priorities. *Biology & Philosophy 9,* 167–195.
143. Shrader-Frechette, Kristin, S., & Earl, D. McCoy. (1994c). The Community Concept in Community Ecology. *Perspectives on Science 2(4),* 455–475.
144. Shrader-Frechette, Kristin, S., & Earl, D. McCoy. (1994d). How the Tail Wags the Dog: How Value Judgments Determine Ecological Science. *Environmental Values 3,* 107–120.
145. Shrader-Frechette, Kristin, S., & Earl, D. McCoy. (1994e). What Ecology Can Do for Environmental Management. *Journal of Environmental Management 41,* 293–307.
146. Shrader-Frechette, Kristin, S., & Earl, D. McCoy. (1995a). Defining

Natural Communities and Natural Ecosystems. *Natural Areas Report 7(3),* 4–5.

147. Shrader-Frechette, Kristin, S., & Earl, D. McCoy. (1995b). Ecology and Environmental Problem Solving. *Environmental Professional 16,* 342–348.
148. Shrader-Frechette, Kristin, S., & Earl, D. McCoy. (1995c). Natural Landscapes, Natural Communities, & Natural Ecosystems. *Forest and Conservation History 39,* 138–142.
149. Shrader-Frechette, Kristin, S., & Earl, D. McCoy. (1999). Molecular Systematics, Ethics, & Biological Decision-Making Under Uncertainty. *Conservation Biology 13,* 1008–1012.
150. Shrader-Frechette, Kristin, S., & Earl, D. McCoy. (2000). Community Ecology, Population Biology, and the Method of Case Studies. In Frank, B. Golley & David, R. Keller, editors, *Science of Synthesis: An Introduction to the Philosophy of Ecology.* University of Georgia Press.
151. Shrader-Frechette, Kristin, S., & Earl, D. McCoy. [forthcoming]. *Values, Underdetermination, and Progress in Ecology.*
152. Simberloff, Daniel. (1980). A Succession of Paradigms in Ecology: Essentialism to Materialism to Probabilism. *Synthese 43,* 3–29.
153. Sismondo, Sergio. (2000). Island Biogeography and the Multiple Domains of Models. *Biology & Philosophy 15(2),* 239–258.
154. Sloep, Peter, B. (1986). Null Hypotheses in Ecology: Towards the Dissolution of a Controversy. In *PSA*, vol. 1, S307–S313.
155. Sloep, Peter, B. (1994). The Impact of `Sustainability' on the Field of Environmental Science. In, G. Skirbekk, editor, *The Notion of Sustainability and its Normative Implications.* Scandinavian University Press.
156. Sterelny, Kim. (2001). The Reality of Ecological Assemblages: A Palaeo-ecological Puzzle. *Biology & Philosophy 16(4),* 437–461.
157. Strong, Donald R Jr. (1982). Null Hypotheses in Ecology. In Esa Saarinen, editor, *Conceptual Issues in Ecology*. Boston: Dordrecht Reidel.
158. Taylor, Peter & A. Blum. (1991). Ecosystems as Circuits: Diagrams and the Limits of Physical Analogies. *Biology & Philosophy 6,* 275–294.

159. Taylor, Peter & Yrjö Haila. (1998). Natural Contradictions: Links between Ecological Science and Environmental Politics. *Science as Culture* 7(4).

160. Taylor, Peter. (1987). The Strategy of Model Building in Ecology, Revisited. *Eighth International Congress on Logic, Methodology, and Philosophy of Science 2,* 308–311.

161. Taylor, Peter. (1989). Revising Models and Generating Theory. *Oikos 54,* 121–126.

162. Taylor, Peter. (1990). Mapping Ecologists' Ecologies of Knowledge. *PSA 2,* S95–S109.

163. Taylor, Peter. (1992). Community. In Evelyn, F. Keller and Elisabeth, A. Lloyd, editors, *Keywords in Evolutionary Biology*. Cambridge: Harvard University Press.

164. Taylor, Peter. (1999). From Natural Selection to Natural Construction to Disciplining Unruly Complexity: The Challenge of Integrating Ecological Dynamics into Evolutionary Theory. In, R. Singh, K. Krimbas, D. Paul, & J. Beatty, editors, *Thinking about Evolution: Historical, Philosophical, and Political Perspectives*. Cambridge: Cambridge University Press.

165. Taylor, Peter. (2000). Socio-Ecological Webs and Sites of Sociality: Levins' Strategy of Model Building Revisited. *Biology & Philosophy 15(2),* 197–210.

166. Taylor, Peter. [forthcoming]a. Apparent Interactions in Community Models.

167. Taylor, Peter. [forthcoming]b. Distributed Agency within Intersecting Ecological, Social, & Scientific Processes.

168. Vepsäläinen, Kari & John, R. Spence. (2000). Generalization in Ecology and Evolutionary Biology: From Hypothesis to Paradigm. *Biology & Philosophy 15(2),* 211–238.

169. Waters, C. Kenneth. (1990). The Conceptual Basis for a Nonequilibrium Theory of Succession. *Trends in Ecology & Evolution 5(4),* 123.

170. Weber, Marcel. (1999). The Aim and Structure of Ecological Theory. *Philosophy of Science 66,* 71–93.

171. Wimsatt, William, C. (1980). Reductionistic Research Strategies and their Basis in the Units of Selection Controversy. In *Esa Saarinen, editor, Conceptual Issues in Ecology.* Boston: Dordrecht Reidel.

172. Wimsatt, William, C. (1994). The Ontology of Complex Systems:

Levels, Perspectives and Causal Thickets. In Robert Ware and Mohan Matthen, editors, *Canadian Journal of Philosophy Supplement 20.* Calgary: University of Calgary Press.

173. Zdravko Stevcic, Rovinj. (1996). Logical Problems of Ecology [Logicki problemi ekologije]. *Filozofska Istrazivanja 16(1),* 19–23.

Webliography

1. Chinadialogue—an independent, non-profit organisation dedicated to promoting a common understanding of China's environmental challenges, based in London, Beijing, and San Francisco
2. Delhi Greens—first environmental blog of India, at the forefront of Indian and global environmental movement. Award-winning, UN endorsed project
3. Dot Earth—environmental blog by Andrew Revkin at NYTimes.com
4. Earther—environmental blog published by Gizmodo Media Group.
5. HuffPost Green—section of the Huffington Post news website, content aggregator and blogs site
6. Inhabitat—founded in 2004 by New York City designer and architecture graduate Jill Fehrenbacher—weblog about trends toward environmental sustainability
7. TreeHugger—acquired by Discovery Communications on August 1, 2007—rated the top sustainability blog of 2007 by Nielsen Netratings
8. Encyclopedia of Earth (EoE)—a component of the Digital Universe project—about the Earth, its natural environments, and their interaction with society
9. Encyclopedia of Life Support Systems (EOLSS)—sponsored by UNESCO—an interdisciplinary encyclopedia, inspired by the sustainable development movement
10. Geo-wiki—founded in 2009 by the project partners the International

Institute for Applied Systems Analysis, University of Applied Sciences Wiener Neustadt and University of Freiburg—to help improve the quality of global land cover maps

11. Global Change Master Directory—part of NASA's Earth Observing System Data and Information System (EOSDIS) and NASA's contribution to the Committee on Earth Observation Satellites (CEOS)—a directory of Earth Science data sets and related tools/services
12. Global Earth Observation System of Systems (GEOSS)—being built by the Group on Earth Observations—to provide near-real-time environmental data, information and analyses
13. Google Earth Engine—developed by Google, in partnership with Carnegie Mellon University and the United States Geological Survey—providing satellite images of environmental change
14. GRASS GIS (Geographic Resources Analysis Support System)—whose development has involved a large number of federal US agencies, universities, and private companies—used in academic and commercial settings around the world, as well as many governmental agencies and many environmental consulting companies
15. Gravity Recovery and Climate Experiment (GRACE)—using two satellites launched from Russia; overall mission managed by the Jet Propulsion Laboratory in California—to study climate and oceanography and geology
16. Green Map—based in New York City—enabling local mapmakers to make environmentally themed maps
17. OneGeology—launched on August 6, 2008, at the 33rd International Geological Congress (IGC) in Oslo, Norway—to enable online access to dynamic digital geological map of the world for everyone
18. Semantic Sensor Web—using standardization efforts of the Open Geospatial Consortium (OGC) and Semantic Web Activity of the World Wide Web Consortium (W3C)—managing data about the global environment
19. UNEP GEO Data Portal—the authoritative source for data sets used by the UNEP and its partners—covering themes like Freshwater, Population, Forests, Emissions, Climate, Disasters, Health and GDP
20. A\J: Alternatives Journal—based in Ontario, Canada, "Canada's Environmental Voice," website and bimonthly magazine

21. Earth Negotiations Bulletin—published by the Reporting Services arm of the International Institute for Sustainable Development—covering negotiations, workshops and conferences on a variety of subjects in environmental policy and international law
22. Ecology Law Currents—produced by students at the UC Berkeley School of Law—an online publication containing "short-form commentary and analysis on timely environmental law and policy issues"[1]
23. Environment and Energy Publishing—based in Washington, D.C.—covering environmental and energy policy and markets
24. Environment News Service—environmental news agency and website—based in Lincoln City, Oregon
25. Environmental Research Letters (ERL)—based in California—publishing peer-reviewed research across the whole of environmental science
26. The Environmentalist—public interest news site covering politics, business, climate, history, lifestyle, world news, science, editorial, links and resources. Carried by major wire services (Reuters, USAToday, Sun-Times Group...).
27. Environmentmagazine.org—hybrid, peer-reviewed magazine and website on "Science and Policy for Sustainable Development," based in Philadelphia
28. Green Builder Media—based in Eastsound, Washington, publishes the North American monthly business-to-business sustainability and green construction magazine Green Builder Magazine.
29. Green Matters—environmental news website based in New York City
30. Grist—online, non-profit environmental magazine and blog(s)—based in Seattle, Washington—"A beacon in the smog"
31. The Oil Drum—published by the Institute for the Study of Energy and Our Future—analysis and discussion of energy and its impact on society
32. Our World 2.0—online magazine created by the United Nations University—based in Tokyo, Japan—focuses on climate change, peak oil, biodiversity and food security
33. Sarawak Report—online investigative journalism news site focusing on deforestation and corruption in Sarawak, Malaysia

34. Yahoo! Green—based in California—gathering news, ideas, and discussion
35. Green.TV—supported by UNEP—broadband TV channel for films about environmental issues
36. How to Boil a Frog—managed by Jon Cooksey—a humor-oriented website with free short videos
37. The Mother Nature Network—founded by Chuck Leavell and Joel Babbitt—eight channels: Earth Matters, Transportation, Business, Technology, Food, Home, Lifestyle, and Family
38. BlueLink Information Network—based in Bulgaria—offering services (in Bulgarian and English) related to the environment
39. Ecojesuit—published by the Global Ignatian Advocacy Network-Ecology (GIAN-Ecology) of the Society of Jesus in Brussels and Manila
40. EnviroLink Network—based in Pittsburgh, Pennsylvania, USA—a clearinghouse for environmental information on the Internet, with chat rooms, bulletin boards, electronic mailing lists, and a worldwide forum
41. Environment & Society Portal—the Rachel Carson Center's open-access digital archive and publication platform. It makes environmental humanities research accessible to academic communities and the interested public worldwide, reaching more than one million users. In addition to its peer-reviewed born-digital publications (Arcadia and Virtual Exhibitions) and the Rachel Carson Center's journal RCC Perspectives, the portal curates scholarly and popular materials in its Multimedia Library.
42. GEOLibrary—materials posted from various governmental and non-profit institutions—occupational safety and health and environmental health information
43. GLOBE Program—operated by the University Corporation for Atmospheric Research (UCAR) out of Boulder, Colorado—a worldwide, hands-on, primary and secondary school-based education and science program
44. Greenlifestyle—originally a mailing list in June 2007—with an ongoing focus on the exchange of practical and cost-effective green living tips and information that can be applied in Indonesia

45. National Environmental Information Exchange Network—managed by states, tribes, and the U.S. Environmental Protection Agency—exchanging environmental data
46. National Institute for Environmental eScience (NIEeS)—was a collaboration between Natural Environment Research Council (NERC) and the University of Cambridge—promoting and supporting the use of e-Science and grid technologies within the field of environmental research (not updated since August 2008)
47. Wiser.org—user-generated online community space for the social and environmental movement—maps and connects non-governmental organizations (NGOs), businesses, governments, and individuals addressing global issues
48. Berkeley Earth Surface Temperature (BEST)—managed by a team of scientists—an effort to resolve criticism of the current records of the Earth's surface temperatures by preparing an open database and analysis of these temperatures and temperature trends
49. Carbon Dioxide Information Analysis Center (CDIAC)—organization within the United States Department of Energy—providing global warming data and analysis as it pertains to energy issues
50. Carbonfund.org—based in Silver Spring, Maryland—providing carbon offsetting and greenhouse gas reduction products to individuals, businesses, and organizations
51. Climate Change TV—funded by companies, governments and organisations, and produced by the magazine Responding to Climate Change—the world's first web channel specific to climate change videos
52. Climateprediction.net—maintained primarily by Oxford University in England—a distributed computing project
53. CLIWOC (climatological database for the world's oceans)—funded by the European Union—drawing on British, Dutch, French and Spanish ships' logbook records for the immediate pre-instrumental period, 1750 to 1850
54. CoolCalifornia.org—created by a team of California State agencies, universities and a nonprofit organization—a website for Californians with resources to help them reduce their impact on the environment and combat climate change
55. DeSmogBlog—co-founded by Jim Hoggan, president of a public

relations firm based in Vancouver, British Columbia, Canada—providing information about global warming

56. Earth System Grid (ESG)—funded mainly by the United States Department of Energy—distributing data for the IPCC Fourth Assessment Report and the 2014 IPCC Fifth Assessment Report
57. Global Atmosphere Watch (GAW)—established by the World Meteorological Organization—monitoring trends in the Earth's atmosphere
58. Global Historical Climatology Network (GHCN)—managed by the National Climatic Data Center, Arizona State University and the Carbon Dioxide Information Analysis Center, a database of temperature, precipitation and pressure records
59. How Global Warming Works—developed by Michael Ranney, a professor of cognitive psychology at the University of California, Berkeley—to educate the public on the mechanisms of global warming
60. HURDAT—accessible from the website of the National Hurricane Center—database for all tropical cyclone's in the Atlantic Ocean, Gulf of Mexico and Caribbean Sea, since 1851
61. InsideClimate News—based in Brooklyn, New York—online magazine that "covers clean energy, carbon energy, nuclear energy and environmental science"[2]
62. OneClimate—part of Internet portal uk.OneWorld.net—climate news, social activism and social networking site
63. PRECIS (Providing Regional Climates for Impacts Studies)—required by the United Nations Framework Convention on Climate Change—enabling scientists from the around the globe to run a regional climate model towards carrying our research into climate change
64. RealClimate—hosted by Environmental Media Services—a commentary site by working climate scientists
65. Skeptical Science—created in 2007 by Australian blogger and author John Cook—publishing articles on current events relating to climate science and climate policy, and maintaining a large database of articles analyzing the merit of arguments commonly put forth by those who are skeptical of the mainstream scientific opinion on climate change
66. TerraPass—based in San Francisco—a social enterprise that funds greenhouse gas reduction projects in the United States, and offers carbon offsets to individuals and businesses who wish to reduce their

carbon footprints

67. Watts Up With That?—created in 2006 by California meteorologist Anthony Watts—arguing for climate change denial
68. Quake-Catcher Network—operated by Stanford University and UC Riverside—a distributed computing project using computer-based accelerometers to detect earthquakes
69. ARKive—launched in the UK by Sir David Attenborough—a global initiative to locate and gather films, photographs and audio recordings of the world's species into one centralised digital library, for increased environmental awareness
70. Biosafety Clearing-House—established by the Cartagena Protocol on Biosafety—to facilitate sharing of information on living modified organisms (also known as GMOs)
71. Global Biodiversity Information Facility (GBIF)—whose Secretariat office is hosted by the University of Copenhagen—publishing data provided by many institutions from around the world
72. Global Invasive Species Information Network (GISIN)—global network of organizations—for sharing information about invasive species
73. Greenmuseum.org—nonprofit online museum of environmental art, with a forum, a toolbox, and event listings
74. Alsos Digital Library for Nuclear Issues—part of the United States' National Science Digital Library (NSDL)—a searchable collection of vetted annotations and bibliographic information for resources including books, articles, films, CD-ROMs, and websites pertaining to nuclear topics
75. Global Fishing Watch—initiated by Google in partnership with Oceana and Skytruth—a project to enable citizens to monitor fishing activities globally
76. Global Oceanographic Data Archaeology and Rescue Project (GODAR Project)—initiated by the National Oceanographic Data Center (NODC) and the World Data Center system (WDC)—to provide oceanographic data to climate change and other researchers
77. Global Temperature-Salinity Profile Program (GTSPP)—a cooperative international project—to develop and maintain a global ocean Temperature-Salinity resource

78. World Ocean Database Project (WOD)—established by the World Data Center system (WDC) and the U.S. government's National Oceanographic Data Center (NODC)—intended to centrally store oceanographic data
79. PR Watch—project of the Center for Media and Democracy—having a stated mission to expose deceptive and misleading public relations campaigns, especially about US environmental issues
80. Globe at Night—initially a NASA educational program—crowdsourcing measurements of light pollution in the night sky
81. Microwave News—independent and not aligned with any industry or government agency—reporting on the health and environmental impacts of electromagnetic fields (EMFs) and other types of non-ionizing radiation
82. Clean Energy Project—by the Department of Chemistry and Chemical Biology at Harvard University—a distributed computing project researching organic compounds for solar cells, polymeric membranes for fuel cells, and molecule assembly
83. Global Energy Network Institute (GENI)—non-profit research and education organization—with focus on the interconnection of electric power transmission networks between nations and continents, emphasizing tapping abundant renewable energy resources
84. Wind ENergy Data & Information (WENDI) Gateway—launched by the Environmental Sciences Division of Oak Ridge National Laboratory—providing access to wind energy-related data and information from a wide spectrum of sources
85. Downsizer—set up in October 2004 by the founder members—"a resource for people who want to live more sustainably"
86. Worldchanging (2003–2011)—was an online magazine about sustainability and social innovation
87. EcoCyc—bioinformatics database for the bacterium Escherichia coli K-12
88. Registry of Toxic Effects of Chemical Substances (RTECS)—until 2001 maintained by US National Institute for Occupational Safety and Health (NIOSH)—database of toxicity information

INDEX

C

D

E

F

G

S

T

U

V

W